D0771352

the undergrowth of science

the undergrowth of science

Delusion, self-deception and human frailty

WALTER GRATZER

OXFORD
UNIVERSITY PRESS

Great Clarendon Street, Oxford OX2 6DP

Oxford University Press is a department of the University of Oxford.
It furthers the University's objective of excellence in research, scholarship,
and education by publishing worldwide in

Oxford New York

Athens Auckland Bangkok Bogotá Bombay Buenos Aires Calcutta
Cape Town Dar es Salaam Delhi Florence Hong Kong Istanbul
Karachi Kuala Lumpur Madrid Melbourne Mexico City Mumbai
Nairobi Paris São Paulo Singapore Taipei Tokyo Toronto Warsaw

and associated companies in
Berlin Ibadan

Oxford is a registered trade mark of Oxford University Press
in the UK and in certain other countries

Published in the United States
by Oxford University Press Inc., New York

First published 2000

A catalogue record for this book is available from the British Library

Library of Congress Cataloging in Publication Data
Gratzer, W. B. (Walter Bruno), 1932-
The undergrowth of science: delusion, self-deception and human
frailty/Walter Gratzer. p. cm.
Includes bibliographical references and index.
1. Science—History. 2. Science—Philosophy. I. Title.
Q125.G785 2000 509—dc21 00-036309

ISBN 019 850707 0 (Hbk)

Typeset by EXPO Holdings, Malaysia
Printed in Great Britain by
Biddles Ltd, Guildford & King's Lynn

Contents

Introduction

Science, according to Jacob Bronowski, is founded on an 'explicit social contract between scientists, so that each can depend on the trustworthiness of the rest'. That the system generally works pretty well is not (I hope) in doubt. This is not to say that the contract is not occasionally breached, for science harbours its social delinquents, just like any other professional group. The great nineteenth-century mathematician, Charles Babbage, in his jeremiad about the decline of science, as he saw it, defined three kinds of malfeasance; these he called forging, trimming and cooking. Forging was the outright invention of data, and was thankfully rare. Trimming was the cosmetic 'massaging' of data, so as to display them to best advantage; this, in legal rhetoric, would be called *suggestio falsi*. Cooking meant discreetly losing the data that were out of line or did not help the hypothesis, that is to say *suppressio veri*. Trimming and cooking are not so uncommon, and they merge at one extreme into the exercise of judgement and experience. They almost invariably accompany an inner certainty on the part of the scientist that his conclusions are correct and scarcely need yet more evidence.

There are some grey areas then, but if scientists told each other open lies so frequently that (as in politics, say, or big business) the suspicion that one was being deceived were to prevail, the scientific enterprise would become futile. In any case, reproducibility of results is so fundamental to the scientific process that fraud cannot be long concealed. There have been many excellent treatises on fraud in science; it happens, it is a nuisance and it can lead to personal tragedy for the innocent, but it does not seriously impede progress. I shall not be concerned with it here.

Nor is it my purpose to recount tales of scientific lunacy, of crack-pots and obsessives, those who believe that we live on the inside of a hollow sphere open at the poles, or that the Milky Way is God's daisy chain, or the memory of water can be transmitted by e-mail (but see p. 28). The lunatic fringes have been extensively explored by such masters of the genre as Martin Gardner and James Randi. These stories are the counterparts of the comically bad poems, collected in that incomparable anthology, *The Stuffed Owl.* There is also, however, another anthology, in which are held up for public entertainment the misjudgements of the famous. This has the title *Pegasus Descending*, and is much closer to what I have tried to achieve here.

Science does not proceed by observation and deduction. Darwin famously observed that he could not imagine how anyone could doubt that all observations must be for or against something. The danger always is that scientists become too closely wedded to a hypothesis to view the evidence with a dispassionate eye. Often the engagement is intellectual, more commonly perhaps personal. To admit that something cherished and nurtured over the years as a bequest to posterity is wrong, even an object for mockery, may be demanding too much of a fragile ego. The explanations proliferate and the hypotheses are elaborated in the desperate attempt to absorb the unassimilable. This is one of the marks of what the great American physical chemist, Irving Langmuir, called pathological science.

More remarkable is the way that false theories and imagined phenomena sometimes spread through the scientific community. A kind of mass hysteria, with parallels in the world at large, such as UFO sightings, alien abductions, 'recovered memory' and probably chronic fatigue syndrome, takes possession of a hitherto rational population, like a virus of the intellect. On such occasions scientists in some area of research throw aside, to the amazement of their colleagues, the intellectual constraints that had until then guided their working lives. They become selectively uncritical and intolerant of any unsought evidence. Sometimes such a perversion of the scientific method results from external, especially political, pressures, but at other times it is a spontaneous eruption. The communal derangement may last for months or years, but the end is always as sudden, and often as unaccountable, as the genesis of the effect. It leaves behind smouldering reputations and red faces, and provokes deep introspection and a scratching of heads.

I hope the reader will share my fascination with these strange episodes. I make no claims to original scholarship and I owe what I know mainly to the writings of others. I have added fairly extensive references for a deeper study of the topics of each chapter.

CHAPTER 1

Blondlot and the N-Rays

The affair of the N-rays is the best and most widely analysed example of tribal delusion in the history of science. It has all the elements of drama—pathos and comedy, a theatrical *dénouement*, a high reputation brought low and more than a dash of chauvinism and malice.

The year is 1903 and for physicists these are stirring times. Towards the end of the nineteenth century the field had been pervaded by a general sense that man's understanding of the physical world was approaching completion. At least one university in Britain resolved not to fill its Chair of Experimental Philosophy (as physics was then called) on the grounds that so little remained to be discovered. The complacency was shattered by Wilhelm Röntgen's discovery in 1895 of X-rays, and radioactivity soon followed. These revelations caused a sensation outside physics by their spectacular, seemingly miraculous nature, and within the scientific community by the realisation that a deeper layer of knowledge about the structure of matter was after all waiting to be uncovered.

In France memories of the disastrous war with Prussia thirty years before still festered and many French intellectuals were obsessed by the need to uphold the superiority of their culture. As will appear later, chauvinistic assertions of innate differences between the Gallic mind and those of the plodding Teutons and Anglo-Saxons found widespread expression. A nationalistic fervour engulfed all cultural and technological activities, science included. The generally superior successes of the Germans and the British in physics and chemistry were hard for French scientists to swallow and they were acutely sensitive to what they perceived (often rightly) as the dismissive attitude of their European neighbours. French achievements tended therefore to be met in France with extravagant acclamation and in some quarters with a lack of critical caution. It was against this background that the N-ray episode needs to be judged.

René Prosper Blondlot was a highly respected physicist, born into an academic family—his father was a noted physiologist—and a Professor at the University of Nancy. Nancy was no backwater, for although Paris occupied a commanding position in the sciences, as in all else, determined efforts were made around the turn of the century to strengthen the major provincial centres. A faculty of science had been established in Nancy in 1854 and was greatly invigorated after the Prussian annexation of Alsace in 1870 by an influx from the University of Strasbourg, for a large proportion of the faculty there had opted to remain French and many sought shelter in Nancy. In 1885 a national law was passed to allow subventions to universities from regional and municipal councils, as well as gifts from industry. In Nancy civic pride found expression in the construction of fine new faculty buildings, and the high reputation of science at the University attracted funding from the Belgian industrialist, Ernest Solvay, who chose the city as the site of his new chemical works. A physicist on a

René Prosper Blondlot in the robes of the French Academy of Sciences.

visit from Paris reported that the facilities for research were such as any of his colleagues in the capital might envy.

Nevertheless, for scientists the lure of Paris was irresistible and the better provincial centres, such as Toulouse, Lyon and especially Nancy, were seen by ambitious young savants as stepping stones to the Sorbonne or one of the other venerable institutions in the capital. Even the native sons of Nancy on the faculty, among whom Blondlot was prominent, had nearly all received their training in research, and the patronage that went with it, in Paris.

Blondlot's interests were in electricity. He had produced an outstanding doctoral dissertation under Jules Jamin in 1881 and the great Henri Poincaré had extolled an *experimentum crucis* that Blondlot had performed to confirm Maxwell's theory of radiation.[1] J.J. Thomson in England (soon to discover the electron) described this work as 'exemplary'. Blondlot had been elected a corresponding member of the Académie des Sciences and been awarded two of its most important prizes. A new discovery from such a quarter, no matter how baffling, was not therefore to be lightly dismissed. Moreover the world of science was prepared for the emergence of new kinds of radiation to add to the Röntgen rays and the already familiar alpha-, beta- and gamma-radioactive emissions.

At that time the first question asked about any kind of radiation was whether it had the nature of waves or of particles. The known fundamental particles carried a positive or negative charge and could therefore be deflected out of their trajectory by an electric field. Waves, on the other hand, would not be deflected, but as the study of visible light had shown, they could be polarised, that is to say they could be constrained to oscillate in a plane rather than in three dimensions. (Today light is commonly polarised by passage through a polaroid filter, which subtracts out the component of the wave oscillating in one plane; if the emerging polarised light impinges on a second polaroid with its axis parallel to that of the first all the light is transmitted, but if the axes of the two polaroids are at right-angles to one another no light gets through the second one.) Blondlot set out to perform such an experiment on X-rays. If they were waves, he thought, they might be polarised when they emerged from the X-ray tube and if that proved to be the case their wave character would be conclusively established.

To detect possible polarisation Blondlot set up a spark generator, which he placed in the path of the X-rays. If the X-rays were polarised in the same direction as the spark, jumping between the tips of two wire electrodes, the brightness of the spark would be enhanced by the electromagnetic wave of the X-rays. Blondlot then needed only to vary the angle of the spark gap to see whether there was an orientation at which the spark grew brighter. This indeed was what he observed, but his satisfaction must have been tempered by the result of his next experiment: he allowed the radiation from the X-ray tube to pass through a quartz prism before it reached the spark and was perplexed to find that the light-enhancing rays appeared to be deflected, for it was already established beyond doubt that X-rays were not refracted by such a prism. Blondlot pondered the matter and arrived at what was to prove a calamitous inference: the brightness of the spark, he decided, was being enhanced by something other than the X-rays, which after all would have passed straight through the prism and missed the spark gap. This something could only be another form of electromagnetic radiation, which was to be called N-rays in honour of Blondlot's native city, Nancy. Blondlot, however, was initially cautious; not only did he recognise the danger of self-deception in descrying small differences in brightness, but he also knew that X-rays falling on some materials could produce light, which might affect the background illumination, and indeed the X-ray generator itself gave off a glow. Blondlot therefore devised an arrangement for detecting changes of brightness photographically. The spark-gap detector was enclosed in a cardboard box to exclude visible light, but not, as it transpired, the N-rays. Under the spark gap was a diffusing screen of ground glass and beneath that a photographic plate. N-rays passing through the box and falling on the spark gap increased the blackening of the plate occasioned by the light from the spark. So far, so good. Blondlot now sought to improve the experiment by introducing a screen, opaque to N-rays, between the X-ray source and the cardboard box, so as to allow the rays to fall

"N" RAYS

A COLLECTION OF PAPERS COMMUNICATED TO THE ACADEMY OF SCIENCES

WITH ADDITIONAL NOTES AND INSTRUCTIONS FOR THE CONSTRUCTION OF PHOSPHORESCENT SCREENS

BY

R. BLONDLOT

CORRESPONDENT OF THE INSTITUTE OF FRANCE
PROFESSOR IN THE UNIVERSITY OF NANCY

TRANSLATED BY

J. GARCIN

INGÉNIEUR E.S.E., LICENCIÉ-ÈS-SCIENCES

WITH PHOSPHORESCENT SCREEN AND OTHER ILLUSTRATIONS

LONGMANS, GREEN, AND CO.
39 PATERNOSTER ROW, LONDON
NEW YORK AND BOMBAY
1905

Title page of the English translation of a selection of Blondot's papers on N-rays.

uninterruptedly on the box in the line of the spark gap. Water proved to be opaque to the N-rays and so wet cardboard was chosen to serve as a shield. The box could now be illuminated by the radiation for a fixed time interval, with and without the wet cardboard screen in place, and the change in blackening of the photographic emulsion due to the N-rays could be observed. But then, in the quest for speed and improved sensitivity, Blondlot switched to other means of detection. The most satisfactory was a phosphorescent (calcium sulphide-coated) screen, which glowed when radiation fell on it.

Blondlot conducted an exhaustive series of investigations on the properties and sources of N-rays, the results of which he published in the proceedings of the French Academy of Sciences (*Comptes rendus hebdomadaires des séances de l'Académie des Sciences*, known as the *Comptes rendus* for short) starting in 1903. X-ray tubes were not the only source, for N-rays came also from incandescent filaments (Nernst glowers), the gas mantles used for domestic lighting, red-hot metals and indeed the sun. A platinum wire, heated to a dull red, glowed more brightly where N-rays fell on it, though, as the

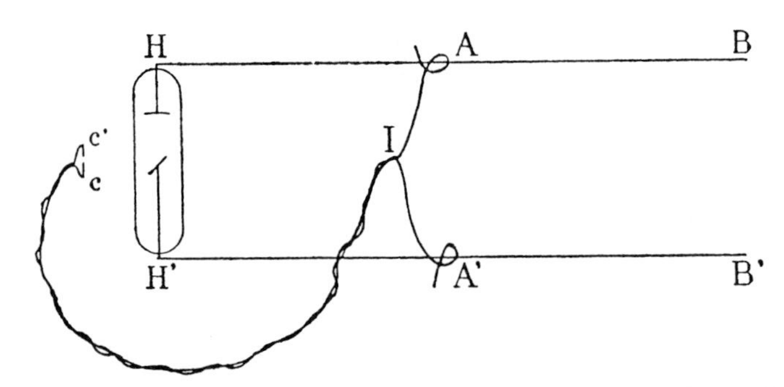

Diagram from one of Blondlot's papers, showing a typical set-up for studies on the properties of N-rays. HH' is a source of N-rays, a gas discharge tube, activated by insulated wires leading from an induction coil, B,B′. This experiment was designed to discover whether the N-rays were polarised. The wires A,A′ lead to a spark gap, operated by the same circuit. The spark gap could be rotated and the intensity was expected to dim or brighten according to whether the spark was aligned along or perpendicularly to the plane of orientation of the N-rays.

Below is shown a photographic image of the spark with the two orientations of the spark gap. This may seem an irreproachably objective measurement, but subjectivity entered into the timing of the exposures.

unchanged electrical resistance showed, without increase in attendant temperature. The Journal, *Nature* commented that the extraordinary properties of the new radiation endowed it with 'unusual interest' and a promise of 'great utility'.

Most materials opaque to visible light, it turned out, were transparent to the rays—paper, wood and thin sheets of metal for example—and Blondlot fashioned prisms and lenses of aluminium to refract and focus the rays. Quartz and mica were also transparent, but water and rock salt were opaque.

The German physicists, who led the world in the study of electromagnetic radiation, were dismayed by these reports. Heinrich Rubens, who was famous for his exploration of the infrared (long wavelength) part of the electromagnetic spectrum, thought that what

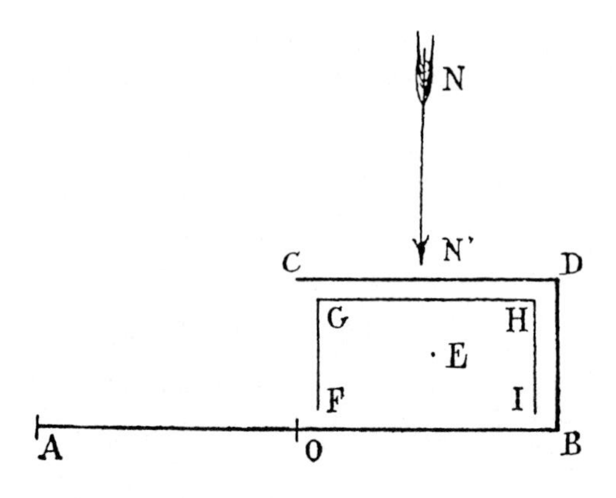

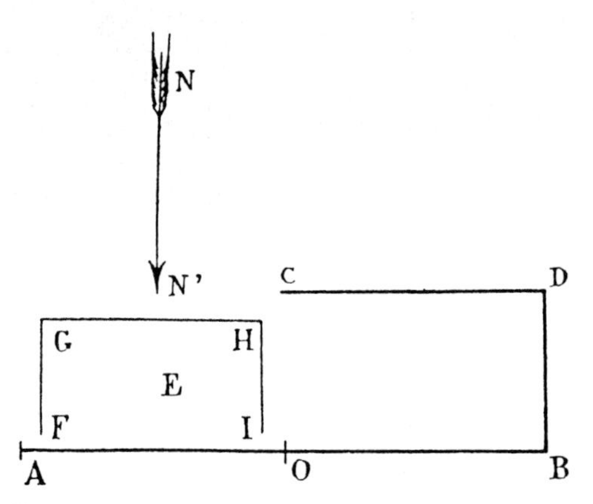

Here is how Blondlot tried to eliminate the dangers of subjectivity in comparing luminosities on phosphorescent screens. The diagram from one of his papers shows a beam of N-rays, NN', falling on a cardboard box, FGHI (supposedly transparent to the rays, though not of course to room light), resting on a photographic plate AOB. A spark is generated inside the box at point E. CD is a lead screen, wrapped in wet paper, opaque to N-rays. Then when the cardboard box is shielded from the N-rays (upper disposition), the N-rays cannot act on the spark, whereas an increase in brightness (blackening of the photographic plate) occurs when the box is not screened (lower disposition).

Blondlot claimed was impossible, a contradiction of Maxwell's theory. Even the longest-wavelength infrared rays were blocked out by aluminium foil thicker than one-tenth of a millimetre (which of course is totally opaque to the shorter-wavelength light of the visible

Below is shown one of Blondlot's published demonstrations of increased plate blackening (right) when N-rays (in this case from two heated iron files) impinged on a spark.

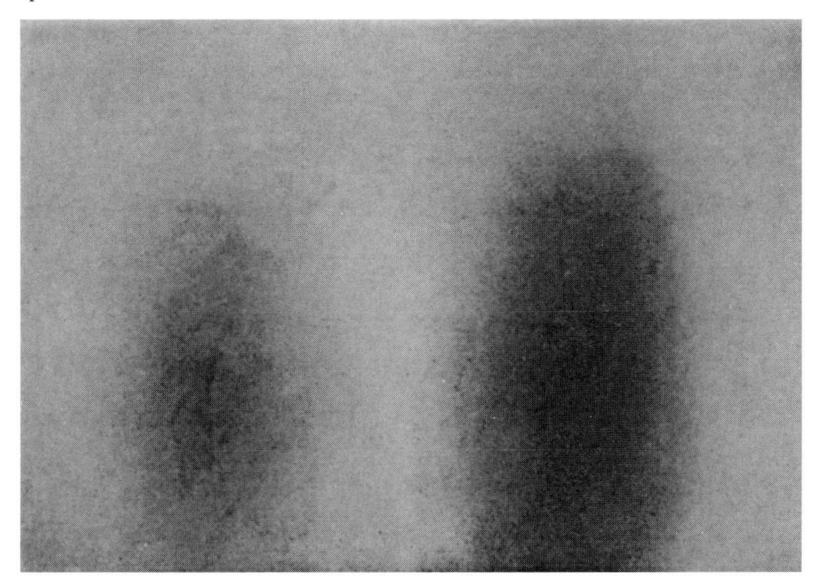

spectrum). Quartz, too, was opaque to long-wavelength infrared. And if Blondlot's measurements of the refractive index of quartz to N-rays were right, their wavelength would have to be at least four times that of Rubens's longest-wavelength infrared radiation. How then could they be transmitted by the quartz? Blondlot responded that there might be 'five octaves' of unexplored radiation between the infrared and the visible. Rubens and other physicists in Germany, though unanimously sceptical, tried to reproduce Blondlot's results and failed.

None of this inhibited the ardour of the French, and Blondlot's discovery was eagerly taken up by researchers in Nancy and beyond. Most sensationally, N-rays were found to emanate from living tissues, both animal and plant: that same year, 1903, Augustin Charpentier, the respected Professor of Biophysics at Nancy, reported in the *Comptes rendus* that muscles and nerves of rabbits and frogs gave off N-rays in abundance (and the animals went on radiating after death).

N-rays, he thought, might actually be used to generate images of the internal organs of patients. Even solutions of enzymes, isolated from living tissue, were emitters. Curare dampened the intensity of N-rays produced by nerves. In Paris the young Jean Becquerel (son of Henri, the discoverer of radioactivity), together with André Broca, Professor of Medical Physics (and the son of Paul Broca, revered as the founder of French anthropology and most of all for his studies on the brain) found that anaesthetics suppressed the emission of N-rays. (A later report by the hapless Becquerel had it that the anaesthetics also suppressed N-ray emission from hot metals.)

The noted Sorbonne physiologist and academician, Arsène d'Arsonval,[2] reported that N-ray emission from the brain—more precisely from the region of the brain known as the 'centre of Broca' (after Broca *père*)—increased when the subject talked and especially during mental exertion. N-rays falling on the eye, Charpentier found, induced visual 'hyperacuity', or heightened perception of objects in dim light: a clock face in a darkened laboratory, for instance, became visible to the investigator after he had irradiated his retinas with N-rays. This was of especial interest to French psychologists, for at this time in France all mental disorders were widely held to have organic causes. Perhaps differences in sensitivity to N-rays in the environment could give a clue to the origins of mental states? Parapsychology, too, was taken very seriously in certain academic quarters, and N-rays were seized on as possible agents of otherwise inexplicable manifestations, such as luminous auras, perceptible only to spiritualists, who believed themselves to be blessed with sensory acuity out of the normal run. A prominent medium of the day, Carl Huter, in fact asserted that he had discovered N-rays years earlier, but his claim was rejected by the Academy. A well-known practice of parapsychologists was to transmit a supposed bodily, or 'odic' force along an iron wire. Perhaps N-rays surged along the wire from one subject to the other?

Arsène d'Arsonval had earlier formulated a vitalist hypothesis that envisaged a kind of force-field around all living beings, giving rise to a distinctive type of radiation; this field, he believed, was characterised by physical constants, which would eventually be determined, 'as for any other form of energy'. The state of mind of many intellectuals in France was thus highly receptive to the idea of a new (and French) radiation with physiological origin and effects.

But fate was already preparing to throw the banana-skin in Professor Blondlot's path. A torrent of publications on N-rays began to flood into the *Comptes rendus*, mainly, though by no means entirely, from Nancy. Certain substances, it was found, could store N-rays and re-emit them later. Some materials emitted N-rays when compressed, so a walking-stick held close to the eyes and bent sensitised the retina to the perception of dim objects. There next appeared another type of radiation—the N_1 rays—which issued from metals and *dimmed* the luminosity of a glowing light source. Jean Becquerel formulated the hypothesis that N-rays resulted from molecular expansion, N_1-rays from contraction. Further thought persuaded him that there were actually three kinds of N-rays, positive, negative and uncharged, by analogy with the alpha-, beta- and gamma-radiation emitted by radioactive substances. The chemist André Colson, who had already determined that N-rays were given off during certain chemical reactions—depending in at least one case on which of two reagents was added to the other—inferred that a chemical condensation was analogous to mechanical compression, which, as we have seen, squeezed out N-rays. The Academy voted another prize to Blondlot, although at least some members of the committee, and especially the chairman, Henri Poincaré, must have experienced a degree of unease, for the Prix Leconte, valued at about five times an annual professorial salary, was awarded for the sum of Blondlot's works, rather than for his discovery of N-rays. Gustave Le Bon, an influential dilettante with some achievements in science, maintained that he had, in advance of Blondlot (in fact just two months after Röntgen's first paper on X-rays), described a new radiation with the alluring name of 'black light', which could only have been N-rays, capable of passing through metals; but a committee of the Academy declined to validate his claim of priority. There were few reports of N-rays outside France; two at least came from Britain and a London instrument-maker advertised an apparatus for the detection of N-rays, price one guinea.

Yet even in France the enthusiasm was not unanimous. Whereas several physicists and physiologists in Paris (besides the many in Nancy) confirmed Blondlot's observations, there were others who failed, most significantly two physicists of international stature, Paul Langevin (whose liaison with Marie Curie after her husband's death was to create a celebrated scandal) and Jean Perrin. Langevin had visited Nancy to observe Blondlot's experimental procedures and had then set up the same experiments in the Collège de France laboratory of one of the elders of the academic establishment and a champion of Blondlot, Éleuthère Mascart, respected for his work in many branches of physics. The results were negative.

In Germany, the physicists Heinrich Rubens, Otto Lummer (a leading expert in the measurement of light intensities) and Paul Drude reported failure to reproduce Blondlot's results, as did Lords Rayleigh and Kelvin and Sir William Crookes in England. Two letters from British researchers appeared in *Nature*, describing in some detail failed efforts to detect N-rays, using phosphorescent screens; the first (from the Cavendish Laboratory in Cambridge) concluded that, since Blondlot must surely have taken all necessary precautions to guard against self-deception, it followed (by inscrutable logic) that he had discovered a radiation to which some men are blind. The second letter suggested a mechanism by which the French observers could have been led astray, namely dilation of the pupil, which was known to occur during mental effort and would increase the perception of brightness; moreover, concentrating on an object in the peripheral visual field engenders a relaxation of the muscles in the eyeball controlling the curvature of the lens. And, the letter sagely concluded, 'might not the mental condition of some observers in a state of expectancy react on the intrinsic muscles of their eyes, and thus they see what they think they should see?' Later, the distinguished

German physicist Peter Pringsheim made an observation on perception at the optical threshold: to test whether you can perceive a white screen (a sheet of paper for instance) in near-darkness you may hold your hand above it and move it back and forth. If you can see the movement you can surely see the screen. But Pringsheim showed that the test worked even if the screen was placed *above* the hand, for though you do not know whether you can really see the screen, you know where your hand is and a willing imagination does the rest.

The Parisian academic establishment—Mascart, Arsène d'Arsonval, Becquerel *père*, André Broca—held firm, and even the aged chemist, Marcellin Berthelot, sometime Minister for Public Instruction and Foreign Minister (who even appears as a character in a novel by Zola), gave his support, but a queasy feeling of doubt began to diffuse through the scientific community. French science was unsure of itself and in perpetual fear of the Germans. So when Claude Bernard, arguably the pre-eminent physiologist of the nineteenth century, died, an obituary notice in the popular journal *Revue Scientifique* contained the following passage: 'Claude Bernard, of all Frenchmen, was the one whom the haughty science of the trans-Rhine could least dare to disregard'. (The translation does not capture the resonance of the original: *Claude Bernard était de tous les français celui qui l'orgueilleuse science d'outre-Rhin osait le moins méconnaître.*) True, three French physicists, Henri Becquerel and Marie and Pierre Curie, had shared the Nobel Prize in 1903, but one, Marie Curie, was not really French and was often exposed to xenophobic obloquy, and her husband had been excluded from the established academic system, by reason of his refusal to conform to its corrupt mores and his unconcealed atheism and socialist convictions. The Sorbonne had expressed its concern that young American scientists, sent by their professors to gain experience in Europe, went to Germany and Britain, seldom to France.

Henri Becquerel's discovery of radioactivity was long ignored in Germany, and an explanation offered by a German physicist received wide publicity: spurious radiations, he thought, notably that described by the aforementioned Gustave Le Bon in the *Comptes rendus*, were seen as a speciality of the French.

In 1904 the *Revue Scientifique*, a sober organ of high quality, solicited, under the title, '*Les Rayons N existent-ils?*' the opinions of a dozen leading French scientists on the question. Louis Cailletet, physicist and pioneer of the liquefaction of gases, owned that, were it not for Blondlot's standing and high achievements, he would think N-rays a delusion: he had attended a public demonstration in Nancy, at which many of the audience, 'notably ladies', had perceived the effects very clearly and 'expressed their pleasure with cries of admiration', but for his part, having stopped his ears and summoned all his concentration, he had seen nothing. Nor had he fared better in his attempts to reproduce the supposed effects in his own laboratory. Yet he had, he continued, as much confidence in Blondlot as in himself, and he fervently hoped, in any event, for the sake of French science that Blondlot had not been deceived. But as for the work of Jean Becquerel, 'who hypnotised—no chloroformed—a five-franc piece', he confessed, though he had no evidence on the matter, that it would take much to persuade him of its truth.

Cailletet was followed by the celebrated inorganic chemist Henri Moissan. 'Do you think', he began, 'scientific questions can be resolved by plebiscite?' He would offer no opinion. M. Moissan, the editors interjected here, did not deny having attempted to reproduce Blondlot's experiments, and he had certainly not reported success. Pressed further, he admitted to having his own opinion, but he preferred to keep it to himself. Nothing he might say would advance the matter like a good experiment: 'let it be done!' Professor Berget of the Sorbonne gave reasons for doubt and added the following anecdote: Heinrich Rubens, one of the world's foremost authorities on electromagnetic radiation, received one day an emissary from Kaiser Wilhelm II. His Majesty had heard speak of the new radiation, discovered in France, and he desired Professor Rubens to prepare a demonstration of the effects of N-rays for the next morning. The experiments seemed simple enough and so Rubens set out on his task with a tranquil mind. But, alas, the phenomena proved elusive and after a night's fruitless toil he had to confess to a mortifying failure, which had occasioned mirth in the Kaiser's court. In the next issue of the *Revue* Rubens denied the story (which was presumably a malicious canard, contrived by Berget). Rubens and his colleagues, Hagen and Lummer (specialists in measurement of light intensities), had indeed striven to repeat Blondlot's experiments, not at the Kaiser's instigation, but to satisfy themselves. It was only after a month of exacting work and upon mature reflection that they had decided to make known their opinion: despite their high esteem for Blondlot, who had given proof of his merit by many fine contributions to diverse areas of physics, they believed with Berget that N-rays did not exist.

A succession of other *savants*, including Pierre Curie and Georges Goüy, all testified that they had been unable to see any of Blondlot's effects. Professor Monoyer of the Faculty of Medicine at the University of Lyon was more forthright: no, he did not believe that N-rays existed. As to Blondlot's argument that the ability to discriminate differences in luminous intensities varied between individuals, this was not in doubt, but was it likely that 'a phalanx of six observers at Nancy would possess retinal sensitivity great enough to see variations in brightness imperceptible to hundreds of other physicists and physiologists in France, England, Germany and Italy?' Only two voices of dissent were raised in the survey: a physiologist at Nancy was in no doubt that he had seen the effects, clear as day, and an associate of the Medical Faculty of Lyon insisted that tempered steel emitted N-rays, as detected photographically. The *Revue Scientifique* restricted itself to the comment that carefully controlled experiments were now more necessary than ever, but the damage was done. In another issue the editors reported morosely that at the 6th International Congress of Physiology in Brussels in September of 1904 the German delegates had absented themselves from the session on N-rays—an open gesture of their contempt. The *Revue* asked only that N-rays should not become a target of ridicule—a stick to belabour French science—to foreigners.

The blow fell in the summer of 1904, delivered by R.W. Wood, Professor of Physics at Johns Hopkins University in Baltimore. Wood had been urged by Rubens to take on Blondlot, and he was the man for the task. A distinguished spectroscopist, who studied the nature of light emitted by different elements and molecules when stimulated, he was also a man of wit and a noted *farceur*. He disseminated faked photographs of space-ships, alarmed the citizens of Baltimore by spitting into puddles on wet days while surreptitiously tossing in a small lump of sodium, which would hiss and explode in a sheet of yellow flame, and he made a sport of exposing spiritualist mediums, then at the peak of their popularity in America, as in Europe. Wood later built the world's most powerful spectrograph for recording visible and ultraviolet spectra and in a paper describing its construction he explained how he cleared cobwebs out of the narrow forty-foot optical tube by training his cat to walk through it. Wood published a book of comic verse, which he also illustrated, entitled *How to Tell the Birds from the Flowers*, intermittently in print to this day. ('The Parrot and the Carrot' for instance, graced by a drawing that makes them appear almost identical, goes like this:

The Parrot and the Carrot you may easily confound,
They're very much alike in looks and similar in sound.
We recognise the Parrot by his clear articulation,
For Carrots are unable to engage in conversation.)

His biographer records that Wood tried, as soon as he read about N-rays, to repeat Blondlot's experiments in his own laboratory. He was clearly sceptical from the outset: 'I attempted to repeat his observations, but failed to confirm them after wasting a whole morning'. He also alluded to another discomfiting incident for French physics, when the eminent Henry S. Rowland in Philadelphia discovered an important new phenomenon, the movement of a charged body in a magnetic field; his observation was confirmed in two German laboratories, one of them Röntgen's, but some years later Victor Crémieu in Paris reported negative results and asserted that Rowland's effect was illusory. A young colleague of Wood's then performed further experiments, which upheld Rowland's conclusions. Poincaré, on behalf of the Academy, thereupon suggested a collaboration between the French and American workers to sort the matter out and the affair ended amidst a good deal of acrimony and embarrassment. Wood now suggested that Blondlot should submit to a similar collaboration, with the clear implication that his data were no more to be relied upon than Crémieu's five years earlier. Prompted by Rubens and others, Wood invited himself to Nancy, where, as he put it, 'the apparently peculiar conditions necessary for the observation of this most elusive form of radiation appear to exist'.

Wood was courteously received by Blondlot and his colleagues. Fluent in French and German, he resolved to conceal his knowledge of their language so that he could secretly listen to the exchanges between the experimenters, and he conversed with Blondlot in German. What followed was a disaster for the French. We owe the only version of what passed in that darkened laboratory in Nancy in August of 1904 to Wood's report in *Nature*. The first experiment that he was shown purported to demonstrate the increase in brightness of a spark on which the N-rays were focused by an aluminium lens. The observer looked through a ground-glass screen which diffused the light of the spark. When the experimenter interposed his hand between the source of the N-rays and the spark gap, the spark was supposed to dim. 'It was claimed', Wood wrote, 'that this was most distinctly noticeable, yet I was unable to detect the slightest change.' His hosts told him that his eyes were not sufficiently sensitive, whereupon he suggested that *he* should be allowed to place his hand in the path of the N-rays and be told when he had done so. He repeatedly put in and withdrew his hand, but 'in no case was a correct answer given, the screen being announced as bright or dark in alternation when my hand was held motionless in the path of the rays, while the fluctuations observed when I moved my hand bore no relation whatever to its movements'.

Blondlot was well aware that observation of small intensity changes by eye was a hazardous procedure—indeed one of his Parisian critics later remarked that Blondlot and his colleagues appeared to feel that awareness of the problem was sufficient to nullify it. But Blondlot, as we have seen, had made efforts to eliminate subjectivity from the observations by using photographic plates to detect intensity changes. He now showed Wood a collection of exposed plates. Wood was not impressed, for, he pointed out in his letter to *Nature*, the intensity of the spark fluctuated quite widely, he thought by perhaps 25 per cent. To allow for long-term changes in intensity, by for example erosion of the electrode wires, the plates had been exposed for accumulated periods of five seconds each, the plate-holder being shifted back and forth by hand between the two positions, that in which the spark was irradiated by the N-rays, the other in which it was shielded from them. Wood was unwilling to accept this evidence, because he did not believe that such obvious changes in intensity as seemingly recorded by the photographic plates would not have been plainly apparent to his eye, and more especially he thought that moving the plate-holder at five-second intervals by hand introduced every prospect of unconscious bias on the part of the experimenter, striving for the 'correct' result. Wood suggested alternative procedures to avoid such risks, based on what we would now call blind trials, in which the observer would not know whether the light source was or was not being exposed to N-rays.

But the decisive moment was yet to come. Blondlot and his assistants had been measuring the N-ray spectrum.[3] The N-rays could be refracted by an aluminium prism and thus, it was claimed, broken down into their component wavelengths, seven in number. The detector was a thin line of a phosphor (calcium sulphide) painted on a piece of card. The card was attached to a curved steel sector, which could be turned by a screw-operated mechanism, fine enough that settings differing in position by only one-tenth of a millimetre could be discriminated. Wood, who had been unable to detect any difference in brightness as he rotated the screw, watched while an observer called out the angles of rotation at which the brightness rose and fell, and in the darkness was able to reach out unobserved and slip the aluminium prism into his trouser pocket.

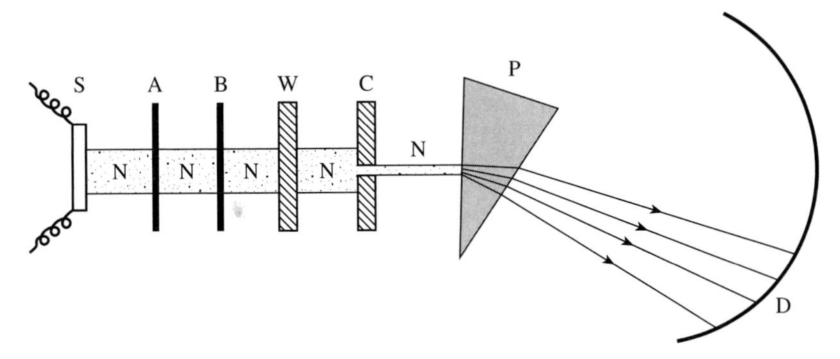

Blondlot's experiment for breaking down N-rays into their separate wavelength components. The source of N-rays (S) was a type of lamp called a Nernst glower (a filament that becomes conductive and incandescent when heated). The N-rays (N) were passed through screens of aluminium foil (A), paper (B) and wood (W) to ensure that all other kinds of radiation (such as visible light) were eliminated. They then traversed a slit in a wet cardboard (opaque) screen (C) to generate a narrow beam, and were refracted by an aluminium prism (P) and detected on emerging by a phosphorescent line painted on cardboard and mounted on a sector (D). This detector was moved round and the angles at which the brightness (to the eye of an observer) increased were recorded. These positions were supposed to correspond to the wavelengths of the constituents of the N-ray spectrum. It was this experiment that caused Blondlot's downfall, when the visitor, R.W. Wood, surreptitiously removed the prism without disturbing the recording of the N-ray spectrum.

This, he reported, in no wise impeded the process of locating the spectral lines.

Wood did not even stop there, for he was also shown Charpentier's experiment, in which an iron file, held close to the eye, enhanced the clarity of a clock dial on the other side of a dimly lit room. Wood saw no such effect and again managed to deceive Blondlot by presenting to his eye a piece of wood the same size as the file, which Blondlot averred increased his visual acuity. Wood's published account and his conclusion were brutal: the experimenters who had seen any of the effects of the supposed rays were 'in some way deluded'.

Wood's letter in *Nature*, translations of which quickly appeared in the *Revue Scientifique* and in the *Physikalische Zeitschrift* in Germany, caused dismay in France. Joseph Le Bel, a pioneer of stereochemistry (the shapes of chemical compounds) wrote of the shame for French science when 'one of its distinguished savants measures the positions of lines of the spectrum while the prism reposes in the pocket of his American colleague'. Blondlot himself, however, did not surrender. He reacted with indignation to Wood's exposé in *Nature*: Wood had tricked him and abused his hospitality. His first response took the form of a letter to the *Scientific American*, in which he stated that the imperfect German in which he had been compelled to communicate with Wood had caused misunderstandings. He enlarged there and elsewhere on the technique for observing brightness differences by eye and he quoted a standard source, Helmoltz's textbook of physiological optics, on the training required by a would-be observer of such effects.

Blondlot's epigones reacted in often absurd ways. One of his assistants during Wood's visit, when taxed with the apparent redundancy of the prism for dispersing the N-rays, replied: 'But that does not surprise me at all: I had already found this at Nancy. The rays deflected by an aluminium lens stayed deflected when one removed the lens.' Chauvinistic reactions also asserted themselves: it was scarcely surprising that the Germans in particular had failed in their attempts to see the optical effects perceived by true Frenchmen, for their sensibilities were inferior and were further blunted by their brutish diet of beer and sauerkraut.

Mascart and others continued to proclaim the reality of N-rays. A noted physicist, who gave his name to an important phenomenon in spectroscopy, Aimé Cotton, and one of his associates, Charles Raveau, visited Nancy in 1906 to see for themselves and were unconvinced. 'By myself', Cotton wrote, 'I observed the expected changes, sometimes very strongly; but when I submitted them to a control [by which he meant a second, independent observer], they disappeared totally'. Gutton, one of Blondlot's most ardent followers, showed Cotton his exposed photographic plates, which proved the reality of N-rays, but again Cotton remained sceptical: one might well get such results, he thought, if one knew all along which side of the plate *should* show the greater intensity. Cotton demanded an independent repetition of this experiment by an observer who did not know what to expect. Gutton curtly refused, but did afterwards collaborate with Blondlot's most powerful supporter in Paris, Mascart, in a series of new experiments, with positive results.

Scientists throughout France now aligned themselves on one side or the other and the *Revue Scientifique* organised a poll. France's most eminent inorganic chemist, Moissan had caustically demanded whether scientific questions were now to be settled by plebiscite. Albert Turpain from the University of Bordeaux made a careful and unsuccessful attempt to repeat a series of Blondlot's and Gutton's observations and submitted his results to the *Comptes rendus* for publication. The editor was Mascart and he rejected the paper. Turpain, enraged, wrote to the *Revue Scientifique* to make public Mascart's reasons for the rejection: 'Your results can be explained simply by supposing that your eyes are insufficiently sensitive to appreciate the phenomena'. 'If', Turpain now declared, 'N-rays can only be observed by rare privileged individuals then they no longer belong to the domain of experiment'. And what of Blondlot himself? He said little more, except to offer the usual explanations for the failure of his various opponents to observe the effects that he had documented in such detail. To various suggestions for collaborative, blind tests, involving such devices as sets of samples in sealed boxes, aired for the most part in the *Revue Scientifique*, Blondlot replied with a letter some months later. 'Permit me', it ran, 'to decline entirely to collaborate in this over-simple experiment; the phenomena are much too delicate for this. Let all form their personal opinions about N-rays, either from their own experiments or from those of others in whom they have confidence.' And with that N-rays disappeared from the scientific literature (and presumably from the undergraduate physics courses taught by Jean Becquerel, Edmond Bouty and probably others).

Blondlot's belief in his discovery seems never to have wavered. Fourteen years after the débâcle he wrote to the Academy as follows: 'I declare that I have never had the smallest doubt concerning the phenomena which I have named N-rays and heavy radiation, and now with all my strength I assert their existence, confirmed by innumerable observations that I have not ceased to make.' In 1909 Blondlot, then sixty, had asked to be relieved of his Chair at the University of Nancy and published no more scientific papers, although he brought out a revised edition of his textbook of thermodynamics. After his death in 1930 he left sealed papers which revealed that he had indeed never ceased to experiment on N-rays. Blondlot was a bachelor; in his will he left a large sum to the city of Nancy to be used for the creation and embellishment of public gardens. If you walk through the University district today you may find yourself in a modest thoroughfare called the rue Blondlot.

How then to explain this extraordinary episode of the imaginary rays? How was self-delusion on such a scale possible within a large, sophisticated and rigorously trained scientific community? How could able and experienced physicists, who knew about the hazards of self-deception in evaluating small physical effects, not have been on their guard? Could they have been blind to the opprobrium and ridicule that would descend on them—and the destruction of reputations that would ensue—if it turned out that they were mistaken?

The N-ray affair is by no means the only recorded outbreak of mass-delusion in science. Indeed its cost in time, money and wrecked reputations and careers now seems trifling, compared to that of the most recent such episode, the cold-fusion fiasco (Chapter 6). What these have in common with other such eruptions is that the majority of the community remained sceptical throughout. The evidence for N-rays did not convince the outstanding French physicists of the day, such as Perrin, Langevin and the Curies, who seem never to have bothered with it. Pierre Curie said he had better uses for his time, when approached after the event by a group of diehard N-ray afficionados for help in reopening the debate. At an early stage he had in fact walked over to the Natural History Museum for a demonstration of N-rays, but saw nothing. His friend Georges Goüy was equally sceptical: Blondlot, he wrote in a letter to Curie, had reported that an old steel knife resembled radium, for it had been emitting N-rays for centuries. 'I would assuredly like to see these N-rays, but I have not succeeded.' In a later exchange N-rays were already a joke; Goüy, who was Professor of Physics in Lyon, quipped that he proposed now 'to undertake research on L-rays, so as to bring glory to our city'. And of course no major figure outside France seems to have taken N-rays seriously at all. Nevertheless, Blondlot carried many highly reputable colleagues with him, such as Mascart, Bouty and Charpentier; indeed some 120 scientists, nearly all in respectable centres of learning, published on N-rays.

Jean Becquerel, young, personable and highly regarded, scion of a distinguished scientific dynasty, who, like his father (the Nobel Laureate, Henri), his grandfather and his uncle before him, came to hold a Chair of Physics at the Musée d'Histoire Naturelle in Paris, was one of the most reckless of Blondlot's followers. Here is how he explained what happened (in a passage quoted by the historian, Mary-Jo Nye): 'With such a method you always see the expected effect when you are persuaded that the rays exist, and you have *a priori* the idea that such an effect can be produced. If you add another observer to control the observation, he sees it equally (provided he is convinced): and if an observer (who is not convinced) sees nothing, you conclude that he does not have sensitive eyes.' Becquerel, moreover, had been drawn into this research 'by a scientist whose previous work inspired only admiration; more, another physicist declared that he had seen the effects, and some of Blondlot's students (who since then have become illustrious) published notes on N-rays. Before such authoritative powers a young man, just graduated from the École des Ponts et Chaussées and never yet having done any research, can be excused somewhat for getting carried away'.

Blondlot's delusions were undoubtedly compounded by the confirmation that they received from his assistants, less trained in critical evaluation of experimental results or disposed to question the professional authority of the *patron*, whom they were no doubt eager to please. Blondlot had trusted these assistants and given them responsibility for making independent observations. He had passed a part of his prize money for earlier researches to a valued assistant, and he acknowledged in print the help he had received with the N-ray experiments from his laboratory technician and machinist, L. Virtz; not only had he contributed valuable suggestions for the construction of apparatus, but he had repeated all measurements, 'a necessary control in delicate researches'. Submissive assistants and juniors have often played a part in perpetuating delusions (and all too often, though almost certainly not here, in perpetration of fraud). There was, besides, a sense of collegial solidarity, which disinclined those on the Nancy faculty to find fault with one of their own, or to side with the condescending Parisians. Many more publications on N-rays emanated from Nancy than from all other centres, nor was a single one of them critical.

Nationalistic fervour and unwillingness to give the Germans and Anglo-Saxons best must inevitably have delayed the final capitulation. French scientists' suspicion of foreigners' motives have remained close to the surface to this day. The most recent upsurge of xenophobia in scientific circles followed the publication in 1988 in the British (but really international) journal *Nature* of a paper by Jacques Benveniste and his colleagues from Paris. They claimed that the physiological activity of certain cells was provoked by a substance present in unimaginably small amount. The most highly active biological agents exert an effect at a concentration amounting to perhaps 10^{12} (one million million) molecules in a litre of solution. The concentration of Benveniste's agent turned out to be less than one molecule per litre, so that most often the action would have been produced in the absence of any causative agent at all. The measured effect was small, and, as usual, it needed special skill on the part of the experimenter to observe it. It came as no surprise that Benveniste's work was supported by a homeopathic organisation. The paper was not rejected, possibly because the incredulous referees could not track down the unquestionable flaws in methodology. But it was published together with a disclaimer by the editor of *Nature*, John

Maddox, who had struck a bargain with Benveniste: he would publish the paper if he could visit Benveniste's laboratory and observe the collecting of data. He would be accompanied on this mission by one scientist (Walter Stewart, whose excessive zeal in sniffing out fraud later brought down on him the wrath of the scientific community) and a magician, James ('The Amazing') Randi, famous for exposing claims of paranormal phenomena.

The team of sleuths may have had in mind R.W. Wood's descent on Blondlot's laboratory eighty years before and hoped for a similar outcome. In the event the *dénouement* was somewhat disappointing, for the error was one of statistical sampling, and not easily represented therefore as a dramatic revelation. (Later investigations suggested that fraud by a research assistant may have been involved.) Benveniste was loud in his complaints at the manner in which the investigation had been conducted and he denounced *Nature*, which, he implied, represented an entrenched scientific establishment, unwilling to countenance ideas that deviated from officially sanctioned truths (the cry, of course, of the crackpot down the ages). He explained the phenomena by a mechanism unknown to physical science—the retention by water of a 'memory' of the molecules that had passed through it. The brouhaha undoubtedly caused embarrassment in France, Benveniste was informed by his superiors in the governmental medical research organisation (INSERM) which employed him that he had disgraced French science and his laboratory was shut down.

But why, it was asked, had *Nature* (an elite journal, which rejects as insufficiently interesting up to 90 per cent of papers offered to it) chosen to publish Benveniste's preposterous effusion? A widespread perception in France was that it could only have been a deliberate device by the old enemy, *Albion perfide*, to make French science look foolish. Here is how the French popular science magazine *Science et Vie* saw it: 'The conclusion is that *Nature* has engaged in an intentional process, that of Francophobia. It is precisely the case that the Anglo-Saxons have no high regard for the French scientists. It is precisely the case that there is in Anglo-Saxon scientific circles a general tendency towards arrogance. It is thus necessary to avoid such faux-pas as the affair of "the memory of water", which the Anglo-Saxon press so openly mocked'.[4] The article then goes on to observe that it would be well to find out how it came to pass that an organisation as respected as INSERM could have allowed one of its members so to mislead it. *Science et Vie* missed a trick when it failed to remark that the British medical journal, *The Lancet*, had published only two years earlier a paper from a homeopathic hospital in Glasgow, reporting the curative effect of grass pollen, diluted, with all the ritual of 'succussion', down to one part in 10^{60}! (Bear in mind that the number of atoms in the solar system is many times smaller than 10^{60}.) The authors offered explanations for the phenomenon no less fantastical than those of Benveniste and his colleagues. But of course one could always argue that *The Lancet* is not *Nature*. We shall see more examples of the impact of nationalistic sensibilities on science later.

Notes

1 James Clerk Maxwell laid the theoretical foundation for the succession of discoveries about light and other radiations that so animated physics in the late nineteenth and early twentieth centuries. Maxwell was a man with no enemies. In him were mingled an unmatched intuitive and mathematical brilliance with personal charm and modesty. His verses retain their place in anthologies of comic verse to this day. Maxwell's electromagnetic theory of radiation stands as one of the pinnacles of achievement in the history of science. Its consummation, the *Treatise on Electricity and Magnetism*, must have had a formative effect on Blondlot and his contemporaries throughout the world of physics. Electromagnetic induction had been discovered by Michael Faraday, who had observed that passing a magnet through a closed loop of wire caused an electric current to flow round the circuit. This led to the principle that a flow of electric charge is induced by a changing

magnetic field in the vicinity of the conductor. Maxwell's intuition told him that the relation between magnetism and electricity should be reciprocal—that a changing electric field would conversly give rise to a magnetic field. A rapid oscillation of charge, as when an alternating current is applied to a conductor, would, Maxwell conjectured, set up a wave-motion that would travel through space like ripples spreading on water. Maxwell calculated from known properties of electricity and magnetism that these waves would propagate at a velocity of some three-hundred million metres per second—the measured speed of light. This prediction led Heinrich Hertz soon afterwards to the discovery of radio waves. Maxwell further surmised that light would likewise have the form of electromagnetic waves, with coupled electric and magnetic oscillations, travelling at the same invariant speed. He developed a rigorous mathematical formulation for the propagation of the electric and magnetic wave components, which subsumed all the characteristics of electromagnetic radiation, such as the nature of polarisation (p. 3). Maxwell died in 1879 at the age of forty-seven.

2 He had been *préparateur*—that is to say scientific assistant—to the great Claude Bernard, had worked with, and later succeeded, Charles-Édouard Brown-Séquard (see Chapter 7) and won fame for his studies on electrical stimulation of muscle. In later life he turned to other interests, especially industrial and medical applications of electricity.

3 By spectrum one means the distribution of wavelengths of radiation emitted or absorbed by an assemblage of atoms or molecules. White light (sunlight, say) is broken down by refraction through a glass prism into its different wavelengths from the shortest (violet) to the longest (red) with the full rainbow spread in between; the light from a sodium lamp, on the other hand, is yellow and nothing but a pair of lines in the yellow part of the spectrum emerges from the prism, while a neon lamp produces

lines in the red. In the same way, radiation emitted or absorbed in other parts of the electromagnetic spectrum, from X-rays at the shortest wavelengths to radio waves at the longest, will have its characteristic spectrum, as supposedly did N-rays from various sources.

4 Benveniste has since placed himself still further beyond the reach of satire. In his new Digital Biology Laboratory he has conducted experiments which show, he says, that 'resonance' signals from biologically active molecules, captured in water, can be transmitted over the Internet. 'Signals' from a molecule—caffeine, say—despatched across the Atlantic, will exert an effect in the receiving laboratory…

CHAPTER 2

Paradigms Enow: Some Mirages of Biology

Gurvich and his mitogenic radiation

The story of mitogenic radiation, which began and (more or less) ended in the Soviet Union, bears a curious similarity to that of the N-rays. It lasted longer, diffused a little further into the world beyond the scientific locale within which it arose, and faded away almost imperceptibly with none of the *éclat* that signalled the demise of the N-rays. Perhaps Gurvich's laboratory was too remote and the political ambience too forbidding to draw any avenging sceptic from the West on a mission of destruction.

Alexander Gavrilovich Gurvich (often spelled Gurwitsch, because his works became known in the West through German translations)

A.G. Gurvich

sprang from an intellectual and artistic Jewish family in the Ukraine. He was sent to study in Germany, qualified as a doctor at the University of Munich in 1897 and became fixated on the biology of the developing embryo, a passion that engaged him for the rest of his working life. He married another Russian biologist and doctor, Lydia, who participated in all of his later research and became his leading apologist. After some years at the Universities of Strasbourg (or Strassburg, as it then was) and Berne, Gurvich returned to Russia with his wife and in 1903 was appointed Professor of Anatomy in a women's college in St Petersburg. There, influenced apparently by a close friend, the theoretical physicist, Leonid Mandelshtam, he made an effort to educate himself—perhaps, as it turned out, with insufficient rigour—in physics and mathematics. Thus equipped, he embarked on the ambitious task of clarifying the physical basis of embryonic growth and development of form. He defined the agenda in a monograph, published in 1904 in German, with the title 'Morphology and Biology of the Cell'.

Gurvich revealed at this time a hostility to scientific fashion and an inclination to distance himself from the established order: in association with some like-minded colleagues, he founded the 'circle of small biologists'. This *groupuscule* saw itself as a counterweight to the 'big biologists', the professors in the most venerable institutions of the land. The entry of Russia into the Great War brought an end to such activities and Gurvich was put to work as an army surgeon in a St Petersburg hospital. The October revolution of 1917 brought in its wake much hardship for the intelligentsia and the bourgeoisie. The 'proletarianisation' of science became declared policy and scientists' salaries were reduced to starvation level. If this was a noble social experiment, as many sympathisers among the intellectuals in the West proclaimed, it was, according to the great Russian physiologist Pavlov, 'one to which I would not subject even a frog'. So in 1918 Gurvich and his family sought refuge in the Crimea, where he found a position as head of the Histology Department at a new university in the agreeable city of Simferopol. Many other scientists had also made their way to the Crimea, where the exigencies of the new order were as yet less oppressive. In Simferopol Gurvich found conditions to his liking and there discovered in 1923 the elusive 'mitogenic' (that is to say cell-derived) radiation.

In some sense Gurvich was a man ahead of his time, for he wanted urgently to understand developmental processes, then commonly regarded as inaccessible to physical or chemical reasoning, at the refined level of molecular events. While in Germany, Gurvich had fallen under the influence of the pioneer of developmental biology, Wilhelm Roux, who believed that the steps in the growth of the embryo could be described in mechanical terms and coined the expression *Entwicklungsmechanik* (developmental mechanics). Roux thought highly of his protégé and, as founder and editor of a new journal, *Archiv für Entwicklungsmechanik*, published several of Gurvich's research papers. But Gurvich's imagination became overheated. He strove after a theory that would explain how Roux's mechanics worked. What governed morphogenesis—the process by which a cluster of cells, formed by division of the fertilised ovum, grows in a purposeful, coordinated manner as the animal (or plant) takes shape? In pondering this question Gurvich hit on the idea of a 'supracellular ordering factor'. This was not a substance: it had rather the character of a 'force-field', a concept borrowed from physics, carrying with it the implication of 'action-at-a-distance', as exerted by electrostatic, magnetic or gravitational forces. During morphogenesis, then, newly formed cells, seeking their preordained positions in the assemblage—the 'dynamically preformed morpha' in Gurvich's jargon—are guided by the mysterious field, the geometrical form of which he mapped out. To direct the orientation of the cells required energy, and this came from metabolism—the consumption of nutrients.

Gurvich soon elaborated his theory by postulating two kinds of underlying interactions between the molecules (proteins and so on) in the embryo: on the one hand there were the kinds of forces between molecules familiar to chemistry and physics, and on the other a fugitive, unstable level of organisation, preserved by the influx of the metabolic energy. Gurvich postulated that at least two factors determined whether a cell in a given position chose to divide: there was a 'possibility factor', which was produced by the cell itself, and there was a 'realisation factor', which came from outside the cell. If the second was what is now known as a growth factor, that is to say an activating chemical substance secreted by cells, how was one to explain the coordinated nature of cell division in a developing organism? How in fact could the *élan vital* be disseminated? Why, by some form of radiation, emitted only by growing constellations of cells. Such radiation, issuing from a developing organism, should reveal itself by speeding up the rate of increase of a growing system placed close to it. The mitogenic radiation was no sooner postulated than found, the sensor a growing onion root.

From then on mitogenic radiation became Gurvich's constant preoccupation and he worked on it, together with his wife, his daughter Anna, and a devoted team of research assistants, almost to the end of his life. As with N-rays, some workers confirmed and embroidered Gurvich's observations, others failed and dismissed them as a figment. Angry refutations of such conclusions flowed from Gurvich's laboratory, but with his retirement and the intervention of the Second World War the passions ebbed, and outside Russia Gurvich was largely forgotten.

In 1924, though, Gurvich had achieved eminence on the national scene. No longer a 'little biologist', he had been called to the Chair of Histology at Moscow University. There, however, he eventually fell foul of the infamous Olga Lepeshinskaya (Chapter 9), who, as an unassailable party veteran, had been forced on him as a research assistant. In 1929 Gurvich was driven out of the university, but his scientific reputation saved him from political disgrace, and he acceded to the directorship of a large department in the highly renowned All-Union Institute for Experimental Medicine in Leningrad. He was permitted to travel and lecture abroad and honours came to him for his work on cancer diagnosis (by—what else?—mitogenic radiation). In 1941 the Second World War engulfed the Soviet Union, and after enduring hunger and deprivation during the siege of Leningrad, the Gurvich family, like many of the privileged class, was evacuated by air to the more salubrious environment of Kazan on the Volga. There Gurvich wrote a new account of his field theory and awaited the end of the war. Reinstated in Leningrad, he resumed work on his radiation until in 1948 Stalin delivered Russian biology into the grasp of Lysenko (Chapter 9). The ageing Gurvich, like so many of his biological confrères, was of a sudden ideologically suspect; he was abruptly dismissed from his departmental directorship and withdrew from academic life. He was now able to expound his theories only to a small group of family, friends and disciples in his flat in Leningrad. He wrote another book (never published) and died at the age of seventy-nine in 1954, his faith in mitogenic radiation, like Blondlot's in his N-rays, apparently unshaken.

Let us now examine how this strange delusion came to impose itself on a competent scientific community. On the face of it, many of Gurvich's experiments, designed to establish the existence of mitogenic radiation and investigate its properties, appeared impeccable. In the earliest studies both the source and detector of the radiation were the growing tips of onion roots, where the cells divide with high frequency. The tip of one root, the emitting source, was directed perpendicularly towards a point close to the tip of the second root, the detector. The rate of cell division was then assessed under a microscope and was found to be perceptibly greater in the exposed region than on the side furthest from the source. When a glass plate was interposed between the two roots the effect vanished: the radiation, it followed, was absorbed by glass. Perhaps then it was short-wavelength ultraviolet, to which glass is opaque? Yes, indeed, for a quartz plate, which is transparent to ultraviolet light, did *not* suppress the mitogenic effect.

Gurvich then tried a second detector—a growing yeast culture, which increased in turbidity as the cells multiplied. Exposed to mitogenic rays, the turbidity, measured by an instrument called a nephelometer, increased more rapidly. Or the cells in a small volume of the suspension could be put on a calibrated microscope slide and counted (the technique often used to assay, for example, the number of red blood cells in the circulation as a test for anaemia). The counting was done 'blind', that is to say the observer was not told which samples had been exposed to the radiation and which not. Later still, the yeast cells were set into a block of agar gel and changes in the amount of light passing through, as the cells divided and the turbidity increased, were measured. The onset of growth in a bacterial culture was another measure. Gurvich also sought and found 'secondary' mitogenic radiation, capable of travelling over large distances, which was stimulated by irradiation of a tissue with the primary emanation. But of course biological specimens, as all the

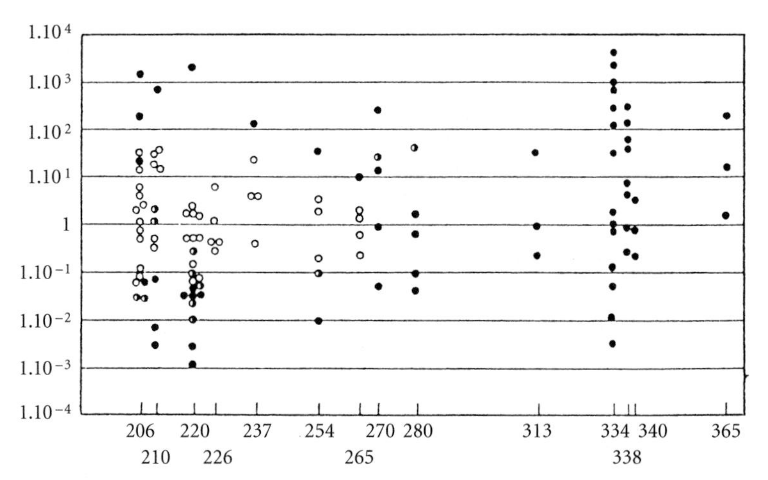

This graph shows the wavelength distribution (horizontal axis with wavelength in nanometres) of individual intensity measurements (with yeast cultures) of emission from a growing onion root. Black circles record a zero effect, open circles a positive effect, and half-filled circles a doubtful outcome.

world knows, are capricious, so if in a series of measurements many failures occurred the experiment was written off as 'inconclusive'—an ominous feature of the analysis.

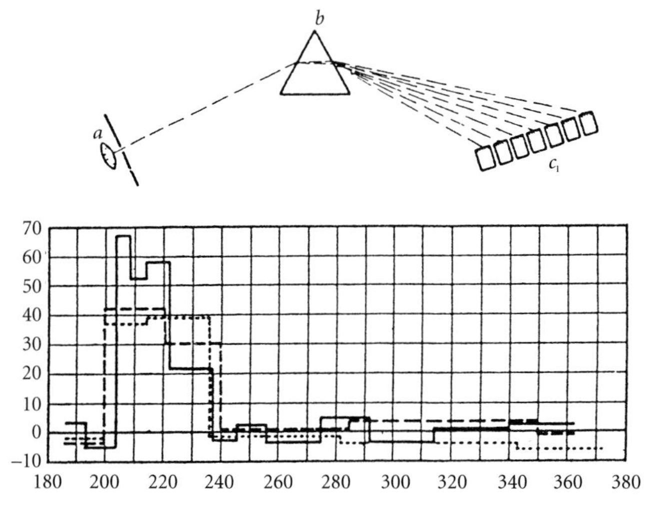

Above: arrangement of a typical experiment to record the spectrum of mitogenic radiation: *a* is a living muscle, supposedly emitting the radiation, which is split into its different components by the quartz prism. The detectors are blocks of growing yeast cells embedded in agar gel, disposed to receive the different wavelengths emerging from the prism. Below is shown the spectrum—the distribution of wavelengths at which the muscle emits. This lies in the short-wavelength ultraviolet region (or would do if it were real). The vertical axis shows the intensities at the wavelength in nanometres marked on the horizontal axis. The different lines are repeat experiments.

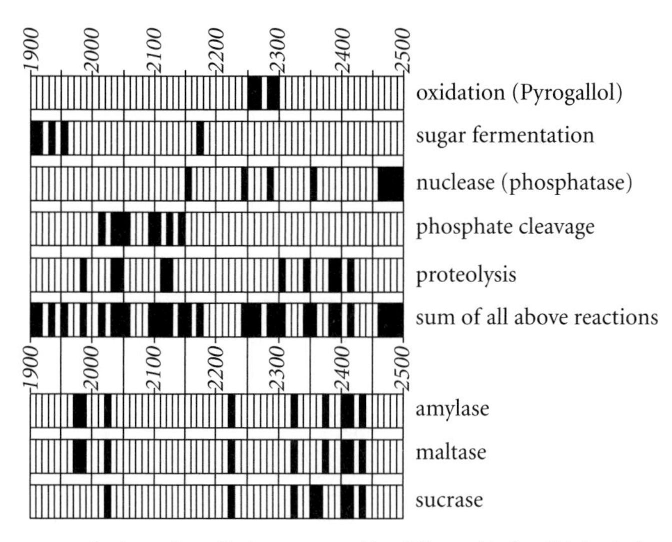

The spectrum of mitogenic radiation generated by different kinds of biological reactions. The boxes represent agar strips, containing growing yeast cells, placed so as to catch different wavelengths emerging from the prism in an arrangement such as that in the picture above. In a tissue in which all of these reactions were occurring the spectrum recorded was the sum of all of these (sixth spectrum from the top). Wavelengths in this case are given in Ångström units (1 nm is 10 Å). Black indicates a positive outcome.

Some among the camp-followers, who plunged so eagerly into the quicksands, later achieved respectability and even distinction in research. The biophysicist, G.M. Frank, for instance, measured the wavelength distribution of the emission from contracting muscle, by passing the mitogenic rays through a quartz prism and detecting the dispersed components with an array of yeast-agar blocks mounted on a sector. The radiation, he found, was confined to the short-wavelength ultraviolet (between 200 and 240 nanometres—visible light extends from about 400 to 600 nm—with maximum intensity at 220 nm). The rate of turbidity changes in the yeast cultures went up and down according as they were exposed to or shielded from the mitogenic emission. The reported augmentation was around 20 per cent, but the total variability between the levels in the 'light' and 'dark' phases was almost as great. Frank ploughed on: he interposed a rotating sector—in effect a rapidly alternating shutter—between a live, twitching muscle and a yeast detector, and found that the muscle emitted mitogenic radiation only in the intervals between the rapid contractions. A resting muscle, which was not working, also beamed radiation—but at a different wavelength.

Mitogenic radiation came out of the test-tube as well, when protein in solution was broken down by a digestive enzyme. This light had components at two wavelengths, 205 and 235 nm. The eye of a living rabbit was a copious emitter, but when the animal was starved the intensity fell off. After five days the radiation could no longer be detected, but on the sixth day it started again, this time, however, at a different wavelength, characteristic of enzymic breakdown of tissues. The blood of starved rats gave off no radiation, nor that of asphyxiated frogs. Potassium cyanide or the anaesthetic, chloral hydrate, quenched the emission from blood.

Other animals too suffered in vain: tumours implanted in mice emitted more strongly than the normal tissue, and continued to emit when the mouse was decapitated. The excised tumour did not emit, but when it was placed in a growth medium containing glucose the radiation reasserted itself vigorously: glucose then was a special nutrient for the cancer cells. This was taken as evidence in favour of a controversial theory of cancer, promulgated by the great German biochemist, Otto Warburg, and known now to have been quite wrong. Excited nerves gave off two kinds of mitogenic radiation, the first associated with oxidative processes, the second with glycolysis, or sugar breakdown, as manifested in the tumour cells. Mitogenic radiation also made incursions into embryology: sea urchin eggs produced a surge of intensity on fertilisation, though, according to another researcher, nothing happened until thirty minutes after the event. Nerves produced characteristic spectra in different regions. Anna Gurvich found that when a frog's eyes were irradiated with different colours, the mitogenic emission switched from one area of the brain to another. The regeneration of an axolotl's amputated tail was accompanied by the onset of mitogenic emission. And so it went: those who sought with sufficient faith and diligence were rewarded.

Exposure to mitogenic radiation, it later turned out, could lead to abnormal development, for it produced deformed sea-urchin larvae. But the culmination came with Gurvich's own report that cancer could be diagnosed by analysis of mitogenic radiation emitted from the blood of the animal or patient. This, he said, preceded all other early signs of the disease. The test was enthusiastically taken up in clinical laboratories throughout Russia. For Gurvich, fame and the Stalin Prize followed.

Interest quickly spread to clinical laboratories (by tradition the least critical and most credulous) in several European countries. The three kinds of mitogenic radiation, distinguished by their wavelengths and resulting from glycolysis, enzymic digestion processes and oxidation, were confirmed, and experiments were devised to show that inorganic oxidative reactions too gave off the third kind (though even so innocuous a process as dissolving salt in water produced a burst of radiation). Carcinogens—cancer-inducing substances such as coal tar—gave off the radiation continuously. White blood cells, with their high metabolic activity, were a powerful source. Blood groups could be identified by the intensity of emissions when different blood samples were mixed. Certain fats also *absorbed* mitogenic rays very strongly and were destroyed on long exposure. Distinctive increases in intensity of emission were seen in hyperthyroidism and in the blood of catatonic and schizophrenic patients. Wounds, while healing, emitted fiercely. Injection of hormones into the pituitary gland caused a life-giving surge.

Mitogenic radiation from the blood faded in old age, in fatigue and in cancer. Senile patients, transfused with the blood of young donors, evinced renewed emission of mitogenic rays, in parallel with clinical marks of 'rejuvenation'. When cancer patients were given radiotherapy the radiation returned (though in Italy patients and animals with tumours gave off *more* mitogenic radiation). Urine also gave off radiation, which was suppressed in cancer. Professor Siebert of Munich reported that enfeebled mitogenic emission was a sure sign of cancer. In such cases a diligent search for a tumour, that had not yet revealed itself by clinical criteria, *always* brought success. (One cannot but wonder how many useless operations were performed to search for and cut out imagined tumours.) And then a Dutch group isolated from blood a substance which they called 'cytogenin'; when injected into healthy subjects, this increased the intensity of radiation from the blood, and it relieved the effects of anaemia and re-started the activity of the blood of octogenarians and cancer patients.

But the frenzy of discovery notwithstanding, criticism—all, it seems, with one exception, from outside the Soviet Union—would not be stilled, and even some of the Russian workers must have felt a little uncertain, for to physicists and chemists it would have gone against the grain of their training to rely on the rate of cell division in an onion root for quantitative measurements. In any event, there were well-established methods for detecting ultraviolet radiation (if this indeed was what mitogenic rays were). The argument of course centred on the exceptional sensitivity of the biological sensor, and Gurvich himself, who was more cautious than most of his followers, mistrusted the physical methods of detection that were on offer. One of these relied on the blackening of a photographic plate, exposed to the radiation. The extent of blackening in general was minute, but at least one publication showed a picture, strikingly reminiscent of the blackening effect produced by N-rays (Chapter 1). A physicist used a Geiger counter (which detects the alpha- or beta-particles generated by radioactive disintegrations), fitted with a quartz window to admit the radiation, and modified to contain a photoelectrically responsive target. The output from the Geiger tube was applied to a loudspeaker. This set-up, it was claimed, had extreme sensitivity, being capable of detecting a few photons (the packets of electromagnetic radiation). When exposed to growing onion roots or mouse tumour tissue the detector experienced a rise of up to 25 per cent in photon flux. A German worker devised the oddest physical detection system of all: he prepared colloidal solutions of inorganic substances (consisting of particles so small that, though made up of a substance intrinsically insoluble in water, they remain dispersed). Metallic gold, for example, is favoured for this purpose. When salts are added such suspensions become incipiently unstable and eventually the suspended particles coagulate into an insoluble mass. On exposure to mitogenic radiation from an onion pulp, this process was accelerated.

It was never clear why very low levels of ultraviolet radiation from the commonest sources (arc lamps and such) should not have engendered a mitogenic effect. It was also remarked that there seemed to be no relation between efficacy and time of exposure or intensity. The experimenters groped for answers: the energy could be absorbed and stored, to be radiated later; the radiation was readily polarised (see p. 3) and would only be absorbed by correctly orientated material. The criteria of pathological science (Chapter 3) are plain to see. And observe how the statistics were treated: in the assays on blood samples, for instance, there were 'occasional failures', but these were due to 'poor' yeast cultures, used as detectors. According to Otto Rahn, Professor of Bacteriology at Cornell University and a leading proponent of mitogenic radiation, such failures had 'nothing to do' with experimental error, because, when mitogenic effects *were* seen they were far outside such error limits. Note that zero effects are thus by definition excluded from analysis of the data, and also that there are *no* results in the range between zero and what is counted as positive. And so the power of 'experience' is allowed to prevail over the 'blind' design of the experiment.

Criticisms accumulated; some workers abroad, notably Otto Rahn, upheld Gurvich's effects, but they began to be outnumbered by others who recorded persistent failure. An intrepid Soviet biologist, Moisseieva, criticised Gurvich's most reliable detector, the onion root, on the grounds that the asymmetric pressure from the tube within which all but the growing tip was confined, stimulated cell division on one side; moreover, because it was exposed on one side to more light, its growth was distorted by phototropism—movement towards the light; and further, the Gurviches selected the best roots for experiments that they expected to matter, while less uniform roots were thought to suffice for controls. Nor had Moisseieva been able to discern any emission from blood or from growing yeast. The most exhaustive of the physical studies came in 1933 from two workers at an address that ensured a respectful hearing—the Department of Colloid Science at Cambridge University, then at the height of its considerable prestige. With a highly sensitive Geiger counter they could observe no emission from fertilised sea-urchin eggs at any stage of cell division, from active spermatozoa or from growing yeast. A source of error, they suggested, might be condensation of water vapour or other volatiles from the samples, hard up against the counter window, onto its surface.

The Gurviches issued fierce rebuttals of their critics, first in Anna Gurvich's book, which appeared abroad in 1932 under the title, *Die Mitogenetische Strahlung*, and later in a series of articles. Their critics' experimental technique was questioned, their data analysis impugned. It was, however, 'a remarkable fact' that, so far, 'purely physical methods' of detection seemed 'in most cases' to fail, but of course they were all operating at the very limit of their sensitivity. The biological detectors though, primarily onion roots and yeast cells, were entirely reliable, as numerous publications from Gurvich's laboratory and many others had indicated.

In fairness to Gurvich, it should be said that he was unhappy at the uncritical enthusiasm with which his discovery was taken up. In 1943 he wrote:

> The very wide interest in the discovery of the radiation, which arose after the first publication [from his laboratory in 1923] and continued for about ten years, was unfortunately not based upon solid scientific ground. The success mostly seemed to be an unhealthy manifestation, insofar as most of those who were attracted to the phenomena discovered and studied in our laboratory were a certain kind of light-minded biologists. Some studies which seemed at first glance to give positive results were based upon only a few experiments; therefore they could not stand up to serious criticism and even compromised the real data.

Gurvich was a conscientious, cautious and by all accounts technically able scientist. To mislead lesser men he had first to mislead himself.

Little by little mitogenic radiation began to disappear from the scientific literature. Mentions were appreciably less frequent when science returned to its pre-war level after 1945, and those publications that did appear were by then all from Russia, in Russian-language journals and therefore little noticed outside Eastern Europe. In 1955 the publications mentioned in English-language abstract journals were down to a single paper and by 1957 mitogenic radiation had vanished from the scientific landscape, at least in Western Europe and America. In Russia it appeared to go underground, to surface again at intervals in books, conference reports and journals unknown in the West. One of the proponents of Gurvich's ideas was his grandson, Lev Beloussov, and Anna Gurvich wrote a retrospective defence of mitogenic radiation as late as 1988 and several more papers in the 1990s. In recent years, in fact, mitogenic radiation has enjoyed a small resurgence, possibly due to the search for explanations of supposed paranormal phenomena, for if living things can communicate with each other by transmitting and receiving electromagnetic radiation there might be no need to postulate the existence of forces alien to science. Some Russians indeed appear even now to believe that cells contain a mysterious entity, called 'bioplasm', that is the agent of communication. As the reaction to N-rays (Chapter 1) also showed, where there are strange effects, seemingly incompatible with the accepted order of scientific knowledge of the time, the parapsychologists and other denizens of the anti-rational undergrowth are seldom far behind.

Most recently a group of workers in China have reported communication between neutrophils: these are one type of white blood cell and an important component of the immune system. They will engulf alien bodies, such as bacteria, and destroy them in a process resembling the killing of domestic germs by hydrogen peroxide. This oxidative reaction is accompanied by the emission of a very low level of light, and such a reaction is termed chemiluminescent. The Chinese workers reported that, when it is exposed to these very low levels of emission, another suspension of neutrophils, kept in a separate but apposed optical cell, is in turn stimulated to emit weak light.

It is now known that chemiluminescence accompanies chemical reactions in which free radicals—highly reactive, short-lived and often destructive chemical species—participate. Such processes require the presence of oxygen. The emission of photons at very low levels, both in the visible and the ultraviolet, from living cells is thus a commonplace, and it is easily detected by modern techniques. This phenomenon has been taken by some as a vindication of Gurvich's work. Fastidious researches from an Australian laboratory have established the chemical source of the chemiluminescence, especially from growing yeast cells: it results from the oxidation by oxygen free radicals of lipids (fats). But when the yeast grows under anaerobic conditions, that is to say when oxygen is excluded, there is no chemiluminescence at all. To equate the weak light emission from metabolic oxidative processes with Gurvich's mitogenic radiation is in any case fanciful, for the only means of detection by which Gurvich set any store was the biological effect; his radiation was defined in terms of its supposed influence on cell growth, whence of course its name, and the Australians (and others) found no such effect at all in yeast (one of Gurvich's two preferred detectors) and other systems.

The cult of mitogenic radiation seems to be still alive then in some dark corners in Russian laboratories, and effects of what is now termed 'ultraweak' radiation from cells on the growth of other cells were reported in the 1980s from a university in Poland and another in Germany. The radiation is supposed to carry information from cell to cell, and, it is suggested, from organism to organism. It is claimed to originate at least partly from DNA, but the evidence is elusive. There is also one representative of the cult in a British university. Working with bioluminescent bacteria—organisms that give off intense light, rather like fireflies, by an oxidative reaction—this worker found in two experiments out of five (shades again of Gurvich) that a fungus, growing in a separate compartment but visible to the bacteria through a quartz window, provoked the bacteria to try harder and emit more light.

Why mitogenic radiation retains a vestigial allure in the fastnesses of the East (and, it seems, an island or two of activity closer to home) is not readily explained. Easier perhaps to engage with is the question of what caused interest in Gurvich's radiation to wane when it did (and not earlier). Partly no doubt it was because Gurvich himself—a scientist who in his day deserved to be taken seriously—had withdrawn from the scene, partly because of the passing of a generation of scientists who, despite the fugitive nature of the phenomenon and its obstinate refusal to emerge from the shadowy limits of detectability, had felt committed to its pursuit. More especially, the processes of embryonic growth and development became less mysterious and the search for new phenomena, transcending the familiar principles of physics and chemistry, no longer seemed necessary to explain them; Gurvich himself, ruminating in old age on the vicissitudes that his creation had undergone, conjectured that the biologists of the time just had no place in their scheme of things for mitogenic radiation and therefore, in their blinkered way, simply excluded it from consideration. And finally, a better grasp of the chemistry that goes on in cells brought the realisation that release of high-energy radiation (that in the wavelength range of the far ultraviolet) is not easily reconciled, except within very narrowly defined limits, with the system of reactions that define life. Or perhaps there was merely a collective change in perception as abrupt and unaccountable as the epiphany that began it all.

The curse of the death-ray

Not all imaginary radiation was beneficent. A persistent belief in Europe in earlier times was that menstruating women radiate a noxious essence. It was said that dough which they kneaded would never rise, that food would spoil if they came near and flowers wilt in their hands. In the twentieth century some biologists felt they had the means to search for the source of this peasant wisdom. And so in the 1920s a series of publications on the subject emerged, mainly in the German medical literature. Many (though not all) reported positive results, and attributed them to a *Menstruationsgift*, or menstrual toxin, which in the English-language journals was termed 'menotoxin'.

In 1929 a bacteriologist, by the name of Christiansen, working in a dairy industry laboratory, recorded his observations in a German botanical journal. He had found that from time to time the fermentation starter cultures failed. He had investigated the problem, and having eliminated 'all other possible causes', was left with the conclusion that these episodes coincided with the menstrual periods of the technician who tended the cultures. Plainly the technician's menstrual blood was giving off harmful emanations. Christiansen now went to the root of the matter: he found that the radiation from menstrual blood passed through quartz, but not glass, induced changes in the appearance of yeast and bacteria in cultures, and eventually killed them. The effect was more pronounced in summer than in winter, and was perhaps therefore engendered by sunlight; so it proved, for when women were irradiated by ultraviolet light their menstrual blood became more potent.

Some time later, the aforementioned Otto Rahn at Cornell University was informed by a research student that throughout one year yeast cultures in the laboratory had at intervals failed to grow when prepared by one woman student. This event turned out to be a monthly occurrence, but it almost ceased in the winter. Deeper study revealed that the woman's fingertip, poised over the yeast culture, killed the cells in five minutes. This powerful effect, Rahn conceded, was uncommon, but it could be induced not only by menstruation but by various other afflictions: lethal emanations came from the finger of a man recovering from an eruption of herpes, from another with hypothyroidism and from Rahn himself, while suffering from a sinus infection. Saliva of some people, placed close to a yeast culture, could change the form of the cells, when examined under the microscope, though here Rahn strikes an unwonted note of caution: perhaps the cells were too close to the saliva and perhaps therefore the effect was not due to radiation at all.

But it was the menstrual effect that caught the attention of the press: scientific proof of the Evil Eye, the newspapers reported, had been discovered at Cornell University. Rahn was embarrassed. This was not what he had told the reporters, he complained in his book on biological radiations, published in 1936; and what they had written could not in any case be right because the radiation was effective only over a distance of a few inches.

While Rahn was busy studying the menstrual emanation, in Moscow the egregious Olga Lepeshinskaya (the nemesis, as we have seen, of Alexander Gurvich) was discovering *necrobiotic rays.* The 'stability', according to Lepeshinskaya, of living cells was enhanced by exposure to weak ultraviolet light, while more intense ultraviolet, as all the world already knew, killed them. Now if weak ultraviolet rays stimulate synthesis of the cellular material, it must follow, Lepeshinskaya reasoned, that when an organism died and the contents of its cells broke down, the energy must be re-emitted, again as weak ultraviolet radiation. Sure enough, dying yeast cells caused blackening of silver bromide grains when these were photographically developed. Interposing a gelatin filter blocked this effect, and so the emitted rays had indeed to be in the short-wavelength ultraviolet—like mitogenic radiation, much of which, Lepeshinskaya suggested, could have been no more than a manifestation of *her* rays. In the Soviet Union there was a small effusion of publications on necrobiotic rays, which were supposedly given off when proteins were coagulated and by other biologically destructive processes; but no scientists of any consequence seem to have troubled themselves with the matter, perhaps because Lepeshinskaya's bite was known to spell death (Chapter 9).

Abderhalden and the protective enzymes

Emil Abderhalden, who was born in 1877 in Switzerland and died in 1950, wielded what seems to have been an unwaveringly malign influence over German biochemistry and physiology for several decades. Many discreditable anecdotes adhered to him, such as the following, which seems to encapsulate his character and was recounted at Cambridge in the 1920s to Professor John Edsall of Harvard University. A young biochemist, recently awarded his Ph.D., arrived for a year's stay in Abderhalden's laboratory. He told his host about some work that he had done at his home university and of which he was very proud. Abderhalden listened with interest. '*Wann publizieren Sie, Herr Doktor*?' he asked. Soon, the young man replied, and with that departed for a brief holiday, hiking in the Alps. He had indeed already prepared a draft manuscript, which, having some inkling of his host's proclivities, he locked securely into a drawer of his desk. On his return he found that the lock had been forced, his manuscript removed and his work already in press, bearing Abderhalden's name as first author.

Abderhalden arrived in Berlin in 1902 and entered the laboratory of one of the greatest of all organic chemists, Emil Fischer. Two years later he became a lecturer (*Privatdozent*) in Fischer's department and in 1908 was appointed Professor of General Physiology at the Veterinary University (the Tierärztliche Hochschule). There he continued the work that he had begun under Fischer on the synthesis of peptides, the building-blocks of proteins,[1] and on proteolytic enzymes. These last occur throughout the body, but most familiarly in the digestive system, where they act to break down proteins into small peptides.

Abderhalden's assistants synthesised many peptides and he published innumerable papers, without apparently introducing any important innovations. This work continued when he moved to the University of Halle in 1911, and soon after he was appointed director of a research institute, not yet built, the Kaiser-Wilhelm Institute for physiology. The Kaiser-Wilhelm Gesellschaft was the organisation set up in 1911, under the patronage of the Kaiser, to promote research in science and technology. The plans for the physiology institute were dashed by the outbreak of war in 1914 and Abderhalden continued his work in Halle, concentrating now mainly on a new class of proteolytic enzymes, which he had discovered in 1909, the *Schutzfermente*, or protective enzymes. Much later these became the *Abwehrfermente*, defence enzymes. Abderhalden flaunted his discovery in a book on the subject, running to many editions.

In essence the thesis was that exposure of an animal to a foreign protein stimulated production of a protective enzyme, which would destroy the protein. It was already known by this time that introduction of foreign bodies, such as bacteria, into the bloodstream induced the production of neutralising molecules, or antibodies, although their identity was only discovered much later. Nevertheless, Karl Landsteiner, an Austrian who, like Abderhalden, had spent time in Emil Fischer's laboratory, had discovered blood groups and so had begun to define the difference between 'self' and 'non-self' and the way in which the body reacts against the latter. Abderhalden's thinking may then have been influenced by these revelations. At all events, he believed he had found the protective enzymes in the blood of animals after injection of foreign proteins, and also in the blood of pregnant animals and women, which had been infiltrated by the proteins of the placenta. Here then was a test for pregnancy, and gynaecologists around the world took note. The procedure was simple: the test material was the coagulated protein mixture in a boiled placenta. When treated with the blood plasma of a pregnant woman the proteins were broken down to peptides, which could be easily identified by a standard colour-generating reaction. From 1912 onwards papers began to appear in the medical literature, in the main extolling the virtues of the method. A poll of women's hospital directors, published in a medical journal, confirmed the favourable impression: none reported failure.

There was one dissenting voice, but this was the most authoritative. Leonor Michaelis was a heavyweight among flyweights, a biochemist whose many contributions, especially to the study of enzymes, secured him an honourable place in the history of the subject. Possibly as a result of anti-Semitism, he had found difficulty in establishing himself in the German academic system and in 1913 was working, full of resentment, in a hospital laboratory. The hospital director asked him to look into the reliability of Abderhalden's pregnancy test. The results were negative. Michaelis visited Abderhalden's laboratory in Halle, but learned nothing to make him change his mind, and so in 1914 the first seriously critical paper was published in the leading German medical journal. Abderhalden was furious and thereafter seems to have done everything in his considerable power to curtail Michaelis's career in Germany. (After much travail and many disappointments Michaelis eventually found his way to America, but never received full recognition for his accomplishments in his lifetime.)

Later reports of failure to reproduce the Abderhalden test came from outside Germany, most notably from Donald D. van Slyke at the Rockefeller Institute in New York. Michaelis was at that time conducting an animated correspondence with a famous German biochemist, by then also at the Rockefeller Institute, Jacques Loeb (portrayed in the character of Professor Gottlieb in Sinclair Lewis's novel of science, set in the Rockefeller, alias McGurk, Institute,

Arrowsmith). Ute Deichmann and Benno Müller-Hill, in their absorbing account of the story of Abderhalden's enzymes, quote a letter to Michaelis from Loeb, dated 1920, in which he writes 'Nobody speaks of the Abderhalden reaction any more in the United States and I am very much surprised to see that in his journal Abderhalden still continues that myth'. Michaelis replies bitterly: 'For me his [Abderhalden's] work is disgusting. My position in Germany has suffered because of my opinion against his pregnancy test. There may be many who see through him, but nobody dares to say anything against him'. Loeb was also scornful of Abderhalden for embracing a theory, propounded by Wolfgang Ostwald, second-rate son of the eminent physical chemist, Wilhelm Ostwald, that physiological processes were controlled by colloids (p. 39). Biological colloids are giant molecules, such as proteins or carbohydrates, but Ostwald did not believe in large molecules, which he held to be clumps of small molecules. He described colloids as a unique state of matter, which did not obey the established laws of physical chemistry. Ostwald was generally held to be a crank, but Abderhalden proclaimed the validity of the colloid theory in his text-book. (Ostwald's own book on colloids was translated into English by his intimate friend, Martin Fischer, a former student of Jacques Loeb and Professor of Physiology at the University of Cincinnati. Several quacks, led by Fischer, much to Loeb's disgust, developed and profitably practised a therapeutic system, which they called 'colloid healing'. Fischer had difficulty getting his papers into print in American journals, and, according to Loeb, complained that he was being persecuted.) 'Abderhalden', Loeb wrote to Michaelis, 'is a psychological puzzle to me, but he is certainly just as big a fool as Wolfgang Ostwald.'

Abderhalden maintained that, with his pregnancy test in use in so many laboratories, it could scarcely be a delusion. The critics had simply failed to get it to work properly, for, he conceded, it could be quite tricky. He and others made modifications to the procedure to render it more reliable, and finally he was gratified to discover that the protective enzymes appeared not only in the blood but also in the urine, where indeed the specificity was higher. This allowed a great simplification to the procedure and the test marched on.

Meanwhile the imaginary enzymes had penetrated into other areas of biology and medicine. Tumours were found to induce appearance of the enzymes, but, more alarmingly, they surfaced in a whole range of mental states. Insane patients were full of the enzymes, which were especially abundant in schizophrenia (or *dementia praecox*, as it was then called). A reliable test could be performed, using the testicle of a bull or a cow's ovary, instead of a placenta as the source of assay material. Different results were found in tissue from different parts of the brain and were correlated with the type of disorder—schizophrenia and Parkinson's disease, for example. The use of the test was advocated in forensic medicine. It was not, said one paper, 'diagnostic for abortion', but it was 'helpful'. Experiments began on injecting patients with protective enzyme preparations to treat various conditions.

Most disturbing, as Deichmann and Müller-Hill discovered, was the plan by the sinister anthropologist Otmar von Verschuer (Chapter 10) and his protégé, Josef Mengele, to study the protective enzymes of different races. These were to be procured from inmates of Auschwitz, deliberately infected with various diseases, and it was agreed that von Verschuer should send an assistant to Abderhalden's laboratory to learn the experimental techniques. An experienced biochemist, who had already worked on the enzymes, was recruited and samples began to arrive from Jews, from Gypsies, from families and from twins, under Dr Mengele's care in the Auschwitz extermination camp.

Wolfgang Ostwald, whose ideas had been championed by Abderhalden, was, for that matter, also a wholehearted Nazi and rabid anti-Semite, and, probably mainly for that reason, loathed Loeb and Michaelis (both Jews). Ostwald's long clamour for recognition in the form of a professorial chair was rewarded only after he joined the Party and Hitler came to power in 1933. In a letter to Martin Fischer he exulted: 'In such a revolution foreign bodies, like

Jews etc. will be disposed of; it is part of the definition of the concept'. As official Nazi Party arbitrator for academic affairs he was able to conduct a vendetta against another enemy, Hermann Staudinger, the leader in the field of giant molecules, who was also a pacifist and considered a political deviant. The feud ended only with Ostwald's death at sixty in 1943. (Staudinger had been able to hold off Ostwald and another creature of the Party, the existential philosopher, Martin Heidegger, Rector of Freiburg University, where Staudinger held a chair, only because of the importance for the state of his work on synthetic rubber.)

It is remarkable how many reputable biochemists, all of them German, participated in work on evaluation and exploitation of Abderhalden's enzymes. Presumably they did not obtain positive results, but perhaps they felt too unsure of themselves to make a pronouncement—for, again, a small difference in enzymic activity between samples of blood or other tissues, which always show some activity anyway, was always apt to be a threshold phenomenon. Deichmann and Müller-Hill believe that it was more likely to have been a consequence of a fear, rooted in the German culture, of speaking out against established authority. A conference on protective enzymes was held in Germany in 1947, only two years after the end of the war. The chairman was the leading German organic chemist of the day, and Nobel Laureate, Adolf Butenandt. It was asserted at the meeting that specific protective enzymes had indeed been found, but Butenandt replied that before such a claim could be upheld the pure substances would have to be isolated and characterised. Abderhalden continued to peddle his enzymes to the end of his life. Deichmann and Müller-Hill discovered that a German biochemist, wanting to work with Abderhalden in order to get to the bottom of the problem, explained how he had succeeded once or twice in making the test work, but had then been unable to reproduce it. Why, Abderhalden asked him, had he wanted to repeat a test that had once worked perfectly well? Dismayed, the younger man departed, having recognised Abderhalden as a fraud.

Abderhalden died, laden with honours, in 1950, but his son, Rudolf, a doctor, kept the faith. The Abderhalden test would assist in diagnostic procedures for optimising *Frischzellen* therapy—the system of rejuvenating treatment practised by the Swiss quack, Paul Niehans (Chapter 7).

A review of the Abderhalden enzymes appeared in 1958 in an American medical journal. It contained 139 references. But the end was nigh and the subject finally disappeared from the scientific literature in 1961. No retraction has ever appeared from any of the laboratories in Germany that published with such profligacy on the subject.

The case of the amorous toad

The Theory of Evolution had, for the more enlightened of Darwin's contemporaries, an irresistible logic and simplicity, which left little room for the views of Lamarck—not that these had ever gained broad acceptance. Jean Baptiste Lamarck was a French naturalist, who first gave the name 'biology' to his pursuit and was widely respected, not least by Darwin, for his classification of plant and animal species according to structural criteria. But his palaeontological studies persuaded him that living creatures can change their form in response to environmental pressures, and that the changes can be inherited by their offspring. His favoured example was the giraffe, a kind of antelope with a long neck. When vegetation was scarce the animal would strive to gather in the leaves on the topmost branches. In time its neck would stretch and its distorted form would be passed on to the next generation, which might then have to exert itself in a similar way. Another instance, Lamarck suggested, was the foot of a bird, forced to live in an aquatic habitat: the bird would spread its feet while swimming and so, as generation succeeded generation, a web would develop.

This theory—the inheritance of acquired characteristics—never exercised a more than marginal influence. Nevertheless, one of the founders of modern genetics, August Weismann, tried to test it by cutting the tails off generations of mice to see whether tailless mice would eventuate. It had perhaps not occurred to him that the hypothesis would also predict that Jewish and Muslim babies would be born with no foreskins. (Julian Huxley illustrated the principle by quoting from *Hamlet*: 'There's a divinity that shapes our ends, / Rough-hew them how we will'.)

Nevertheless, in certain circles the Lamarckian theory retained a tenacious appeal, of which traces remain even to this day. In particular, it was invested with political significance by the founders of Marxism, for it implied that man and society were not at the mercy of nature, as the theory of natural selection implied, but could control their own destinies. The mischief to which this doctrine gave rise will be discussed later (Chapter 9), and it almost certainly played a part in the rise and tragic downfall of Paul Kammerer.

Kammerer was a member of the zoological institute of the University of Vienna, the Vivarium, at the beginning of the twentieth century. He was young, handsome, clever and much sought after in the glittering *beau monde* of the Imperial City. Such was Kammerer's reputation in intellectual circles that the dazzling Alma Mahler, the *femme fatale* who captivated the intelligentsia and married three of its leaders, Gustav Mahler, Franz Werfel and Walter Gropius, worked for a time as a research assistant in his laboratory. Kammerer had strong socialist leanings, and his political convictions cannot but have influenced his approach to biology, which was that of a convinced Lamarckian.

His work began to attract wide attention with the publication in 1909 of a study on adaptive changes in the salamander. Kammerer experimented with two species, the black salamander, which is viviparous, that is to say gives birth to live offspring, and the yellow-spotted, egg-laying, or oviparous species. Kammerer reared animals of the latter kind in a black environment, when, it was known, they tended to lose their markings. Their offspring, Kammerer noted, were mainly black, except for a row of yellow spots along the line of the back. When this second generation was kept on a yellow background the yellow spots dilated and fused into a stripe. The yellow stripe in fact occurs frequently in nature. Descendants of the yellow-adapted animals, kept on a black background, were less yellow, but still much yellower, Kammerer asserted, than the young of parents which had seen only black surroundings.

To reinforce the inference that an acquired characteristic had passed to the offspring, Kammerer now transplanted the ovaries of yellow-spotted female salamanders into brightly striped surrogate mothers. The offspring, as expected, showed the characteristics of the natural mother, *but*, when the surrogate mother was a laboratory-adapted striped individual, the young then reflected the character—spotted or striped—of the father. So, Kammerer concluded, it must follow that these characteristics were inherited through other cells than the gametes.[2]

Kammerer, a resourceful biologist, performed experiments on other animals, to the same end; he claimed to have shown, for example, that severing the trunk-like syphon of the sea-squirt caused the creature to regenerate a longer syphon and to give rise to offspring with long syphons. But his next major undertaking, which brought him fame, or notoriety, and was to prove his undoing, was the study of *Alytes obstetricans*, the midwife toad. This creature normally lived on land, but could also breed in water. The male toads, confined to a watery environment, developed rough, black-pigmented swellings, or rugosities, on their hands. These 'nuptial pads' helped them to cling to the slippery backs of their consorts while mating. Kammerer took eggs from the females and placed them in water. A very few developed and from the resulting adults he selected again the eggs that survived in water. After breeding several generations in this manner he reported that the males which emerged entered the world already equipped with nuptial pads.

The Great War interrupted Kammerer's work and by the time his experiments resumed the climate of biology had shifted. In 1923 he delivered a lecture in Cambridge, which was ill-received. William

Bateson, a leading geneticist and an aggressive academic bruiser, attacked the reported observations in *Nature.* The supposed rugosities, he stated, were in the wrong place. Midwife toads, like frogs, clutched the female from behind by their thumbs, the palms turned outwards. In any case, the supposedly well-developed differences between the palms of the normal and adapted toads were not evident to him, or to others at Kammerer's lecture. Bateson hinted at deception.

A long debate ensued, until in 1926 Kammerer gave the only laboratory-adapted animal that remained to the American zoologist Kingsley Noble to dissect. Noble found no well-developed rugosities and the blackened areas on the palms exuded a black dye, which was plainly not the natural pigment, melanin. It had, in fact, all the characteristics of Indian ink, with which the animal had clearly been injected. Later in the same year, on the night of 23rd September, Kammerer climbed the path up the Hochschneeberg outside Vienna and there shot himself. He had shortly before accepted a high position at the Soviet Academy of Sciences in Moscow, where his political views and his Lamarckian genetics had found favour. He left a letter denying responsibility for the fraud, which someone else, unknown, had evidently perpetrated. He could not, he felt, begin again the long and arduous breeding experiments, which alone could vindicate his claims. It became known that Kammerer was depressed over complications in his personal life, but his patron, the director of the Vivarium, Karl Przibram, accused Kammerer's detractors of causing the tragedy. It is entirely possible that the experiments with the *Alytes* had given the results that Kammerer described, for they were in reality no proof of Lamarckian inheritance. Many animals and plants have cryptic characteristics, known as atavisms, which emerge only under environmental pressure. The toads could have a dormant gene for the formation of nuptial pads, which expresses itself only under a defined environmantal stimulus. Kammerer had in any event selected for this characteristic by rejecting the overwhelming proportion of eggs that did not survive in water. He was aware of the existence of atavisms and, in describing his salamander experiments, gave his own reasons for rejecting this explanation.

The discrediting in the eyes of biologists of Kammerer's most clear-cut experiments—those on *Alytes*—and his admission that a fraud had been perpetrated did not put a stop to loud claims of successful demonstrations of inheritance of acquired characteristics. Soviet Russia presents, as will emerge (Chapter 9), a special case, but in Western Europe several diehards persevered. Prominent among these was the unregenerate E.W. MacBride, Professor of Zoology in the University of London, who proclaimed the primacy of Lamarckian inheritance (and, for that matter, spontaneous generation of life) until the 1940s. It was only with the discovery of DNA replication and the identification of genes that it became apparent why the chances of inserting new (environmentally derived) information into the genome (the total chromosomal make-up of the individual) were so remote. It is indeed known that damage to the DNA of the germ cell, as for instance by radiation, can in principle affect the offspring, but the mechanism of inheritance is now broadly understood and the ghosts of the past have, for most practical purposes at least, been laid to rest. (Once in a while though, as related in Chapter 7, they walk again.)

The experiments of Kammerer and others followed a familiar pattern: who was to judge whether a row of yellow spots had altogether fused into a continuous line or the palm of the toad had swollen and darkened? There were no measurements and no numbers. 'If you can measure it and express it in figures, then', according to Lord Kelvin, 'it is science.' A paper on inheritance from a French laboratory spoke of the duck-like '*comportement*' of a chicken. How was this to be gauged? Like the increased luminosity of a screen when N-rays fell on it, the differences between two animals were evident when there was a predisposition to believe in them. And against this high intelligence and learning were not proof. For an entertaining, if misguided, attempt to rehabilitate Kammerer and Lamarckian inheritance generally, Arthur Koestler's book, *The Case of the Midwife Toad*, is worth reading.

Memory transfer, or eat your mathematics

In the late 1950s the notion gained ground among neuropsychologists that memory was encrypted in the sequences of chemical units in macromolecules. Their thinking was probably influenced by the discoveries then emerging about the nature of the genetic code. Was the information stored in the brain, then, contained in a protein or in a nucleic acid sequence? The nucleic acid, RNA, seemed an especially promising candidate, for its metabolic turnover was already known to be very rapid, and the brain is full of it. It was already apparent by this time that genes were no more than stretches of the chromosomal DNA and that each gene was defined by a unique sequence along the DNA of the four bases, adenine, thymine, guanine and cytidine (A, T, G and C). Their sequence, it was surmised, was translated into a sequence of amino acids, making up the chain of a protein. Not long after, it became clear that an intermediate in this process of translation was an RNA, called the messenger, into which the sequence of the DNA was transcribed.[3]

If memory was indeed enshrined in the sequence of units in a macromolecule then perhaps, if a simple creature could be taught some trick, a new version of the macromolecule would appear in its brain and could be tested for memory function in an animal that had not been trained. First to report success with such an experiment was James V. McConnell, working at the University of Michigan. The organism that he chose was the planarian, or flatworm, *Dugeria dorotocephala.* These primitive creatures live in muddy soil, and know little or nothing, except how to feed and reproduce. McConnell had discovered a paper published by a Dutchman, in Dutch, in 1920, which purported to show that planarians could be made to learn and remember the path to a source of food. Ten years later a group of German researchers described more elaborate behavioural observations on the worms, and McConnell was further encouraged by an American study reporting movement responses to external stimuli. With a colleague, Robert

Thompson, McConnell set out to train planarian worms by standard Pavlovian conditioning methods: they allowed a worm to swim through a narrow water-filled channel and subjected it to an electric shock, preceded by a flash of light. The worm responded to the shock with a vigorous twitch and after this conditioning had been repeated some 150 times the experimenters convinced themselves that the worm had learned to make an avoidance response—the spasm—when subjected to the light without the shock; or rather, the creature was twice as likely to twitch in response to the light flash than an untrained worm. In later experiments the worms were taught to enter one channel rather than another, in which they would encounter an electric shock. The trained worms then changed course when exposed just to the light. But worms would from time to time change course or twitch anyway when travelling undisturbed, so to judge when an avoidance response had occurred required experience and skill on the part of the experimenter, who had, for instance, to divine when a worm was in a placid mood and would not twitch, without the stimulus.

Now the planarian worm is noted for its remarkable capacity to regenerate when cut in two. In a matter of days the head grows a new tail and the tail a new head, complete with brain and eyes. McConnell found that not only the head, but to an only slightly lesser extent the tail, retained the knowledge that the worm had so laboriously been taught. Clearly then the memory substance had distributed itself between the two halves of the worm.

At this point McConnell learned about the work of Holger Hydén in Sweden. Hydén had trained rats to climb wires—a demanding feat—and by a technical *tour de force* he had analysed the base composition of RNA from single cells in the learning centres of the brain. There were differences between the RNA from trained and untrained animals, but the changes were small, the experimental methods stretched to their limit. The work was received with more scepticism than warmth, but it sufficed to convince McConnell that RNA must indeed be the memory molecule.

While he was still pondering the implications and trying to formulate an experimental strategy, two other worm-runners, Corning and John at the University of Rochester, performed what appeared a decisive experiment: they cut trained worms in two and soaked the halves in a solution of ribonuclease, an enzyme that destroys RNA. The outcome was (to those who believed it) dramatic: the heads remembered, the tails forgot, so memory RNA was destroyed by the enzyme before it could establish itself in the newly generated brain.

There were now several groups of worm-contemplators around the United States, and McConnell started a new journal for the dissemination of this distinctly controversial line of research, *The Worm Runner's Digest.* From one of the enthusiasts he learned something new: if the planarians went hungry for any length of time they turned to cannibalism. An inspiration now came to McConnell: suppose a trained worm was eaten by an untrained cannibal. Might the memory molecule not then find its way into the diner's brain? McConnell convinced himself, and proffered statistical evidence to persuade others, that the experiment had worked. When the trained worms were chopped up and fed to their friends, these took in the knowledge along with the meat; or at least, those that had been fed trained worms were more likely by some 50 per cent to make an avoidance response to light than those that had ingested untrained, or as the jargon had it, 'naïve' worms. Later McConnell prepared a crude RNA extract from the minced worms and found that this carried the memory. These results caused much excitement in the community of neuroscientists, and a certain amount of ribaldry elsewhere: there was talk of feeding the professor, or gobbets of his brain to his students or of eating one's physics or mathematics each day at breakfast.

The work was taken up in several laboratories, some of which reported success, an increasing number failure. McConnell spoke expansively of determining the sequence of bases in an RNA encrypting a given memory, then synthesising it in the laboratory—capturing a memory in the test-tube, so to speak—and injecting it directly into the nerve cell, but there were no further claims of new triumphs. As researchers began to lose confidence, the faithful came up with excuses for failure: there were many species of planaria, not easily distinguished, so perhaps the pessimists were working with a refractory kind; perhaps the worms had been given the wrong food or the laboratory environment was unfavourable; perhaps the wavelength distribution in the light flash mattered. But above all else came the experimenter's skill, his empathy with the worms: some people simply did not have the knack. The arguments are by now familiar and predictable.

Then a series of publications began to flow from laboratories in the United States and at least one in Europe on transferred memory in rodents. In one study mice were habituated to a sound—a hammer striking metal—which had initially caused a 'startle response'. When, that is to say, the creatures had become accustomed to the sound they no longer jumped each time they heard it. RNA extracted from the brains of these mice, when injected into the abdominal cavity of untrained mice, suppressed their startle reaction.

In another painstaking series of experiments rats learned to cross their cage to the food dispenser whenever they heard a click. RNA extracts from the brains of these trained animals were injected into the abdomens of their 'naïve' companions. In blind trials (p. 34) the number of times that these recipients showed interest in the food dispenser on hearing a click was recorded; the criterion was whether their nose approached within some defined distance of the feeding cup. Two independent judges agreed in their scoring in 370 out of 375 trials, seemingly a statistically impregnable result. The next publication from the same laboratory reported data on two populations of trained rats: the first group were taught to come for food when they heard a click, the second when they saw a flash of light. Recipient rats inherited the memory that went with the brain RNA: they responded to the click or flash, but not both; the two judges, not knowing which rats were which, achieved agreement in 790 assays out of 800—a conclusive result, one might say, for the probability that such an outcome could occur by chance was less than one in a thousand. Others obtained like results. Yet workers in other leading laboratories found no transfer of memory at all. Moreover RNA is a very labile substance, easily destroyed by digestive enzymes. An experiment with radioactive RNA, which should have been done long before, revealed that none of the RNA injected into the abdomen of a rat reached its brain. Finally, a paper under the names of twenty-three scientists from six leading laboratories reported failure to detect memory transfer in any of a series of independent studies on different species of animals by the various methods that had produced such triumphantly positive answers at the outset.

What could have been the reasons for the illusory successes? The matter was never analysed, but the subliminal effect of an observer, eager for the right answer, is the most likely explanation. In Germany at the beginning of the twentieth century there appeared a horse that could count, tell the time and perform simple arithmetic. This equine prodigy, Clever Hans by name, was examined by a committee of academics and others, who judged its feats of intellect to be perfectly genuine. Its master would, for instance, ask it to indicate which day of the week it was, and the horse would stamp its foot once for a Monday, twice for Tuesday, and so on. Or, being shown a watch, the horse would tell the time in similar manner. More searching tests revealed of course that the animal was reacting to imperceptible movements by its master, who communicated—unconsciously, as everyone agreed—when the correct number had been reached. As will appear, similar signals between experimenters were shown to underlie spurious observations in what seemed to be rigorously controlled experiments in physics (Chapter 3).

The most spectacular evidence for transfer of memory began to emerge in 1968 and continued for some five years. Rats, which are nocturnal animals and shun the light, would normally, when invited to choose, enter a dark rather than a brightly lit chamber. George Ungar and his colleagues at Baylor University in Texas taught their rats to do the opposite, by administering a sharp electric shock whenever the animals followed their natural inclinations. Once trained, the rats were killed, their brains excised and extracts were injected into the brains of untrained animals; these now showed a preference for the light. Mice, too, would acquire the light response when injected with the rat brain extract, so the memory substance was not species-specific. The press pounced on these remarkable revelations and stories appeared in newspapers around the world. The experiments, Ungar asserted, 'are easily and rapidly reproducible and yield unequivocal results which clearly demonstrate the possibility of a purely chemical transfer of some types of acquired information'. Nevertheless, by no means all laboratories found it so easy, and Ungar moderated his claims. 'To succeed,' he wrote, 'too many conditions, most of them unknown, have to be fulfilled, and, if even one of them is neglected, the whole experiment may fail'. Thus, 'the probability of a bad experiment is always greater than that of a successful one.' This is no trivial concession. Ungar, who was undoubtedly determined to get at the truth, then embarked on a collaborative experiment with one of the laboratories (at Stanford University in California) that had signally failed to obtain positive results. Brain extracts were exchanged, but there was no favourable outcome.

Ungar now chose the arduous path of the conscientious scientist: he resolved to purify the memory substance. The closer the preparations came to purity the greater would be the activity and the smaller the statistical uncertainty. Ungar and his colleagues prepared brain extracts from 4000 trained rats and began laboriously to separate the various types of substances that they contained, assaying all the while for potency in inducing the changed behaviour in rats. They found, or so they believed, that the active substance was a peptide—a short chain of linked amino acids, like a small portion of a protein. To be sure, many biologically active substances, including a variety of hormones, are peptides, but this result must not have been pleasing to those who still believed that the universal memory substance was RNA. Ungar and his friends called their peptide scotophobin (that which induces fear of the dark) and they determined the sequence of its fifteen constituent amino acids.

The work, when exposed to view, was adjudged deeply flawed. The paper was submitted for publication in *Nature*, and, because of the exceptional nature of the conclusions (should they prove correct), the editor took the unusual course of publishing it together with the criticisms of the referee, who had been chosen to evaluate the quality of the data. The critique was nearly twice the length of the paper and the study was found wanting in respect both of the behavioural assays and of the chemistry. The amino acid sequence, especially, could not be inferred from the analyses, which revealed that the preparation was highly heterogeneous and contained peptide material with a variety of sequences. For Ungar the next step was obvious: a peptide of fifteen amino acids can be chemically synthesised in the laboratory. When such a pure preparation was tested for activity in several laboratories, none was found.

Ungar apparently did not repine, and moved onto another system, which he believed would allow him to train animals *en masse*, so that larger quantities of memory substance could be recovered for a more convincing chemical analysis: he taught goldfish to discriminate between different colours. Dissection of 17 000 trained goldfish produced 750 grams—nearly two pounds—of brain material. From this he isolated another supposed peptide, or group of peptides, capable of transferring the acquired behaviour to innocent fish. But the world by then had largely lost interest and when Ungar died in 1977 there was no one left to carry the torch. Memory transfer vanished from the scientific literature like a ghost at dawn.

Notes

1 The chemical units that make up a protein are the amino acids. Twenty of these occur in the common natural proteins. When joined together chemically two amino acids form a dipeptide;

addition of a further amino acid generates a tripeptide, and so on. Many short peptides are important in physiology. A protein consists of a polypeptide chain, which can comprise many hundreds of amino acid units. The chemical synthesis of peptides with a defined sequence of the different amino acids was for a long time (and to some extent remains) a preoccupation of organic chemists.

2 The gametes, or germ cells, are the cells of the sperm and ovary, which contain sets of single, rather than paired chromosomes; these chromosomes pair off to produce a zygote, which gives rise to the embryo, containing the genetic material from both parents.

3 The four-letter DNA sequence is copied to make a molecule of RNA, which contains, in a slightly different chemical context, the same four bases, except that the thymine, T, is replaced by the closely related uracil, U. The protein is the product—the manifestation—of the particular gene. There are twenty different kinds of amino acids, all of them specified by triplets of three bases in the DNA sequence. To make the protein, the message (that is the sequence of bases) in the messenger RNA is then scanned by a reading-head (the ribosome) in the machinery of protein synthesis, and the protein, corresponding to the sequence of bases in the RNA, runs off the assembly line. Other elements in the machinery are also made of RNA, in particular much of the complex structure of the ribosome itself, and the adaptors that carry the amino acids to the ribosome for assembly into the protein chain.

CHAPTER 3

Aberrations of Physics: Irving Langmuir Investigates

Capturing electrons

Bergen Davis and Arthur H. Barnes were physics professors at one of America's most venerable seats of learning, Columbia University in New York. Their interest was in the physics of fundamental particles, and the late 1920s and early 1930s, when this story unfolded, saw a great ebullition of new knowledge in this field, due mainly to the work of J.J. Thomson and Ernest Rutherford in Britain, Niels Bohr in Denmark and Arnold Sommerfeld and others in Germany.

The heavy particles of radioactive decay, the α-particles, were known by then to consist of two protons, the positively charged constituents of the atomic nucleus. When these particles were released from a radioactive source in a vacuum tube with electrodes at either end, they travelled in a stream towards the cathode. The other type of charged particle then recognised was the electron; electrons are emitted as cathode rays from a hot cathode in a vacuum tube. They are much smaller than protons and carry an opposite, that is negative, charge. In the prevailing model of the atom, the Bohr–Rutherford atom as it was called, electrons, seen as tiny billiard balls, were in orbit around the nucleus. The electrons were permitted to revolve only in orbits at certain specified distances from the nucleus, for the energies of the electrons, defined by the radii of the planetary orbits, were governed by the quantum theory. The work of Max Planck and Einstein around the turn of the century had established that energy was quantised, that is to say occurred in defined packets. This had profoundly upset most physicists of the time and resistance to this counterintuitive picture did not finally fizzle out for more than thirty years (Chapter 10). But the theory explained experimental observations with extraordinary fidelity. So for instance when an atom absorbs light an electron jumps from an orbit of low energy into one of higher energy; and this energy difference conforms exactly to the energy (measured by the wavelength) of the light that the atom chooses to absorb. And further, an electron that absorbs a sufficiently large packet of energy is liberated from the influence of the nucleus altogether, like a rocket leaving the earth's gravitational field at the velocity of escape. The loss of an electron is known as ionisation. The spectrum of wavelengths of light absorbed on escape of electrons from the various orbits is again precisely predicted.

Now it troubled Davis and Barnes, as it had others, that when α-particles smashed their way through an ionised gas—one containing free electrons liberated from some of the atoms—they were never seen to capture an electron. In 1923 a British physicist called Henderson had performed some experiments that suggested an explanation: the α-particles were ejected from the disintegrating radioactive atom with high velocity and could not therefore be overtaken by the electrons. If the α-particles could be slowed down to a point at which their energy was equal to the difference between that of the free electron and the electron in one of Bohr's orbits, it looked as though capture might indeed occur.

Davis and Barnes resolved to look into the electron-capture mechanism properly. Instead of slowing down the α-particles they would accelerate the electrons. If, to start with, the two beams were travelling at the same speed, would the probability of capture be maximised? Their apparatus was a vacuum tube, inside which was a radioactive source and a hot cathode with a hole, through which the high-energy α-particles would stream. The target was a detector consisting of a zinc sulphide screen, which would register the impact of a

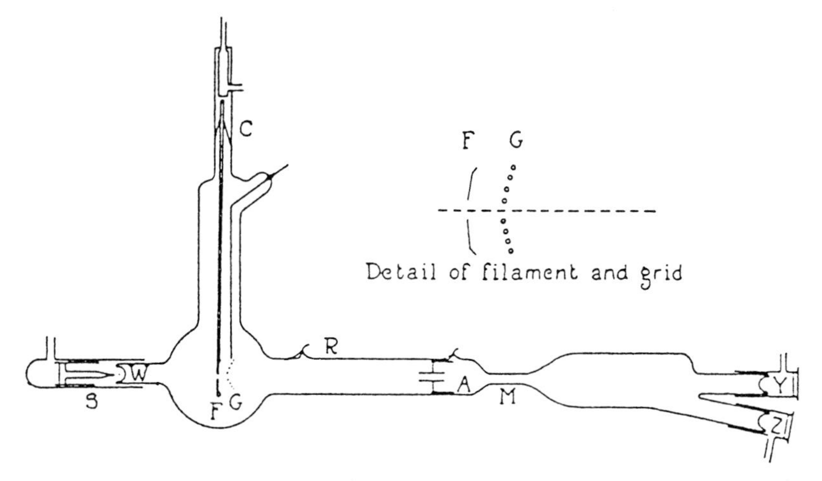

Apparatus designed by Davis and Barnes to study electron capture by α-particles. S was the source of the α-particles, which passed through the window, W, into the vacuum tube. There they travelled straight through to the exit window, Y, or were deflected by a magnetic field, applied at point M, towards window Z. Phosphorescent screens outside the windows detected the arrival of the particles. The electrons, generated by the filament, F, were accelerated to the desired speed by an anode, G, in the form of a grid through which they passed, and directed towards the anode, A, so as to travel along the tube with the α-particles. Science Photo Library.

particle by a small flash of visible light. The scintillations could be counted by observing a small area of the phosphorescent screen through a microscope. A magnetic field could be applied to the tube, which would deflect the α-particles out of their trajectory to strike an appropriately positioned second target. If an α-particle captured an electron one of its two positive charges would be cancelled out, and the singly charged particle would be deflected half as much and would miss the target; so the number of scintillations would be reduced and would give a measure of the fraction of the α-particles that had effected a capture. When the magnetic field was switched off all the α-particles would strike the first screen, while when it was switched on they would be deflected onto the second. The flux was low—some fifty flashes were counted every minute—and, sure enough, when the accelerating voltage was correctly set the number of scintillations decreased. The α-particles were capturing electrons.

Then Davis and Barnes began to explore the effect of changing the energy of the electron beam. Their calculated, and duly observed, accelerating voltage for capture was 590 volts, but now they also detected signals at certain other voltages, and very striking these voltage spectra were. Davis and Barnes had reckoned that the captured electron would enter a Bohr orbit, and, as we have seen, the energy difference between the electron outside the sphere of influence of the nucleus and in each Bohr orbit is known with great accuracy from Bohr's theory. The expectations were borne out: the energies of the peaks detected by Davis and Barnes were those of an electron homing into the different Bohr orbits. The results roused physicists around the world, partly because they carried with them a little theoretical difficulty: the electron, losing part of its energy as it comes in from outer space, so to speak, must, according to Bohr's model, emit half its initial energy as radiation, so if, when it is captured, its energy is only equal to the amount it must have when it has settled into its orbit, there is an energy deficit. The famous German theoretical physicist and expert on atomic spectra, Arnold Sommerfeld, grappled with the problem, as did some others.

Bergen Davis at this point gave a lecture on his research at the General Electric Company's laboratories in New York state. General Electric, like the Bell Telephone Company and some others, had at this time (and for many years to come) the enlightened policy of employing first-rate physicists and chemists and allowing them to do as they pleased, in the expectation that good, if unpredictable, things would flow from their labours. Pre-eminent among them was the great physical chemist, recipient of the Nobel Prize for his work on the properties of surfaces, Irving Langmuir, and it was he who now played the part of R.W. Wood (Chapter 1) in bringing down Davis and Barnes. Langmuir, as we shall see, made a hobby of studying what he called pathological science, and he related the story of his investigations into the subject in a lecture more than twenty years later at the General Electric laboratories.

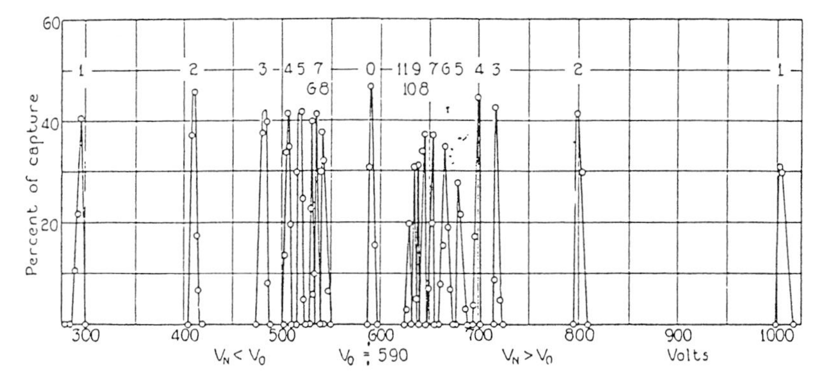

Results recorded by Barnes and Davis: the experimental points (circles) represent the number of scintillations counted, and thus, it was inferred, the number of α-particles striking the screen at each voltage setting. The numbers above the peaks are the quantum numbers, that is to say the rank of the planetary electronic orbit in the atom, 1 referring to the orbit closest to the nucleus (the sun, as it were). The numbers are placed at the energies of the orbital electrons predicted by the theory for the planetary atom.

The first thing that struck Langmuir as he listened to Davis's account of his experiments was that the widths of the peaks, shown in the picture, were no more than about one-hundredth of a volt in a spectrum extending over a range of 700 volts. So when the voltage was tuned on a peak, some 80 per cent of the electrons were captured, while changing it by a mere hairsbreadth caused the probability of capture to drop to zero. In the discussion that followed Davis's lecture, Langmuir asked how long it took to cover the entire 700-volt range if the voltage had to be changed by one-hundredth of a volt at a time and the scintillations had to be counted for several minutes. An experiment at this rate would take some thousands of hours, that is several months. This, Davis replied, was not how the experiment was done, for he and his colleagues had located the spikes in the spectrum by a preliminary search and, since they so precisely followed the order of energies of the Bohr orbits, one knew where to look; thus the exploration could be confined to the immediate vicinity of each line.

Langmuir's suspicions were now fully aroused and over coffee he and his colleagues interrogated Davis further. Here is how the conversation went in Langmuir's words, as transcribed from his famous lecture in 1953, and sounding a little like the dialogue in a Damon Runyon story. Why, for a start, was the efficiency of capture close to 80 per cent at all energies? How did this depend on current density?

> 'That's very interesting,' he said. 'It doesn't depend at all on current density.'
>
> We asked, 'How much could you change the temperature of the cathode?'
>
> 'Well,' he said, 'that's the queer thing about it. You can change it all the way down to room temperature.'
>
> 'Well,' I said, 'then you wouldn't have any electrons.'
>
> 'Oh yes,' he said. 'If you check the Richardson equation and calculate, you'll find you get electrons even at room temperature and those are the ones that are captured.'
>
> 'Well,' I said, 'there wouldn't be enough to combine with all the alpha particles, and besides that, the alpha particles are only there for a short time as they pass through, and the electrons are a long way apart at such low current densities, at 10^{-20} amperes [one ampere divided by ten, twenty times over] or so.'
>
> [Davis] said, 'That seems like quite a great difficulty. But,' he said, 'you see it isn't so bad because we now know that the electrons are waves. So the electron doesn't have to be there at all in order to combine with something. Only the waves have to be there and they can be of low intensity and the quantum theory causes all the electrons to pile in at just the right place where they are needed.' So he saw no difficulty. And so it went.

Now even someone with no education in physics might wonder about the credibility of a professor of physics at a leading university, who argued in such terms. Davis evidently had no doubts or reservations about his experiments. Self-delusion was plainly sapping his critical faculties. Langmuir now decided that he would see for himself and only a few days after Davis's lecture he and a colleague from the Bell Laboratories visited Columbia.

Langmuir, in the same lecture, described what happened. He and his colleague, Clarence Hewlett, sat in the darkened laboratory for half an hour to allow their eyes to become dark-adapted and then they proceeded to count scintillations on the zinc sulphide screen. With α-particles striking the screen, Langmuir counted 50 to 60 hits per minute and Hewlett a few more. When the particles were deflected out of their path by the magnetic field the number fell to about 17—a high background level, Langmuir thought. Arthur Barnes, who was conducting the experiment, counted 230 the first time, 200 the second—far more than his visitors could see—and 25 when the α-particles missed. Langmuir described the scene: Barnes was seated at one end of a long table, gazing at the screen, while his assistant sat at the other end, adjusting the voltage by peering in the gloom at the dial of a voltmeter. The scale on the dial went from zero to 1000 volts, but the assistant could read it with an accuracy, he believed, of at least a hundredth. With the voltage set at a peak position high counts were recorded and when it was set one-hundredth of a volt off-peak the counts fell again. Barnes counted for a standard time of two minutes. Langmuir timed him with a stop-watch and found that the period was in actuality sometimes as short as one minute and twenty seconds.

Langmuir now whispered to the assistant to change the voltage by one-tenth of a volt. The assistant was astonished: such a big change would put it grossly out. Confusion now supervened. Langmuir next asked the assistant to set the voltage and then shut it off in a sequence that he wrote on a card, but the trick did not work, for, as Langmuir soon realised, the assistant would lean forward to set the voltage, but sit back when it was off and needing no fine-adjustment. These movements of course were equally apparent to Barnes, crouched before the phosphorescent screen. Langmuir thereupon whispered instructions to the assistant to simulate, when he set the voltage to a value far from any peak, the effort of concentration required for a precise adjustment. The differences between peaks and troughs in the spectrum vanished.

Excuses followed. As Langmuir recounted it, this was the conversation that ensued:

> 'I said: 'You're through. You're not measuring anything at all. You never *have* measured anything at all.'
>
> 'Well,' he said, 'the tube was gassy. The temperature has changed and therefore the nickel plates must have deformed themselves so that the electrodes are no longer lined up properly.'
>
> 'Well,' I said, 'isn't this the tube in which Davis said he got the same results when the filament was turned off completely?'
>
> 'Oh, yes,' he said, 'but we always made blanks to check ourselves, with and without the voltage on.'

Langmuir was struck by the rapidity with which excuse followed excuse, the unquestioning rejection of any result that did not fit the preconception. 'There is no question but that he is honest,' Langmuir continues. 'He *believes* these things absolutely.'

Langmuir then went to confront Davis with the evidence. Davis was stunned and he could not believe any of it. He wrote soon after to Langmuir that his confidence was not dented, nor did he cancel a presentation of his results before the National Academy of Sciences. To counter Langmuir's evidence he argued that Langmuir and Hewlett counted the very bright flashes on the screen: these, he said, arose from radioactive contamination or some other source. It was the weak (in truth imaginary) flashes, which Barnes had been counting, that signalled the impact of α-particles.

Langmuir now wrote in confidence to Bohr in Copenhagen with his account of Davis's and Barnes's experiment and asked him to forward the warning to Sommerfeld and all others who might be wasting their time on the affair. Work on the phenomenon now stopped, except for a further paper by a British physicist, announcing that he had been unable to reproduce the effect. Not long before, two American workers had also concluded that the peaks of voltage seen by Davis and Barnes were illusory. Their experiments could 'not be explained through spontaneous recombination between electrons and α-particles and must therefore be due to some other mechanism'.

It was nearly a year before Davis finally threw in the towel with a note in the journal *Physical Reviews*, stating that the scintillations on the screen were 'a threshold phenomenon' and that 'the number of counts may be influenced by external suggestion or autosuggestion to the observer'. Langmuir, they wrote, had persuaded them of the inadequacy of their measurements and they had accordingly sought an objective means of counting α-particles, using a Geiger counter and four new X-ray tubes that they had constructed. The results of the visual procedure were not confirmed. The capitulation was not lightly arrived at: the new experiments were laborious and expensive. Davis and Barnes must have salvaged their honour, if not altogether their reputation, in the eyes of the scientific community.

Allison's magneto-optical effect

Irving Langmuir helped to extinguish another celebrated scientific will o' the wisp, the magneto-optical phenomenon of Fred Allison, Professor of Chemistry at the Alabama Polytechnic Institute. In 1927 Allison was studying the Faraday effect, a phenomenon discovered by Michael Faraday in the nineteenth century. Faraday found that the plane of polarisation of polarised light (see p. 3) is rotated when it passes through a liquid placed in a magnetic field. This, and its analogue, the Kerr effect, which is occasioned by an electric field, has been widely used to create rapidly acting light shutters. Allison wanted to measure the time delay in onset of the effect, which would be a measure of how long the molecules of the liquid took to align themselves in the magnetic field.

Allison's set-up is shown in the picture. The liquid is contained in two cells, each surrounded by a coil through which current can be passed to generate a magnetic field. A capacitor discharge generates a spark between two electrodes and at the same time energises the coils. The current directions and therefore the magnetic fields around the two liquid cells are in opposite directions. The light from the spark gap is polarised by the polarising prism and the plane of oscillation

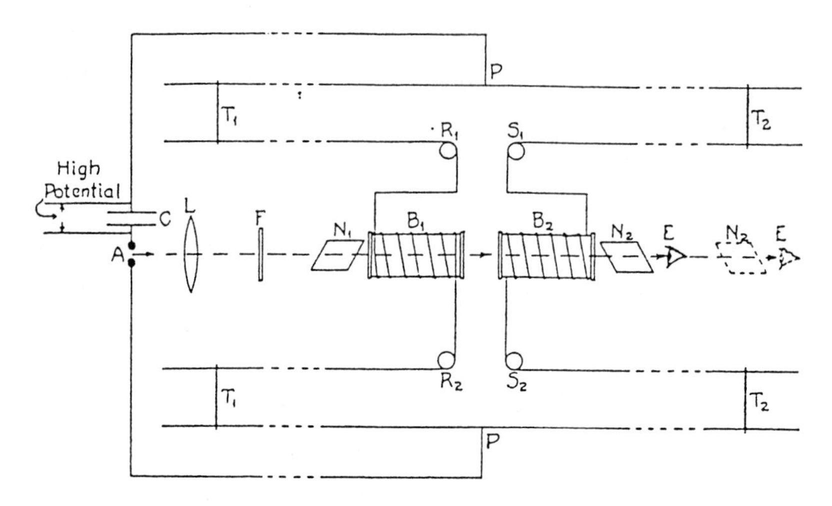

Fred Allison's apparatus for measuring the delay in onset of the Faraday effect, provoked by a magnetic field encircling the sample. The liquid sample is contained in cells B_1 and B_2. The current flows from R_1 to R_2 and S_1 and S_2, inducing a magnetic field. When the capacitor, C, is discharged a spark crosses the gap at A at the same instant as the current surges around the cells. The light from the spark is focused (lens L) and polarised by the polarising prism, N_1. The sample in B_1 rotates the plane of polarisation of the light, and the other cell, B_2, is moved on a trolley, together with the second polarising prism, N_2, to the position at which the rotation of the incident light from the first cell is exactly cancelled out by the counter-rotation in the second (maximum brightness). Here the time required for the light to travel from B_1 to B_2 equalled the delay time of onset of the Faraday effect (the rotation of the plane of polarisation of the light by the magnetic field).

of the light waves is therefore rotated by the liquid—in one direction (clockwise, say) by the first cell, in the opposite direction by the second. The angle of the plane of rotation is measured by the other polarising prism, the analyser, behind the second cell. If there is no rotation and the prisms have their axes aligned, the emergent light will be at its brightest, whereas if their axes are at right-angles (crossed polarisers) the light will be dim. A change in rotation of the analyser leading to such a minimum of brightness will measure the rotation of the plane of polarisation by the liquid. If now the second cell is moved away from the first until there is no rotation, that is to say the opposite rotations engendered by the two cells cancel, then the time taken for the light emerging from the first cell to enter the second cell is the time lag in the onset of the Faraday effect. Since we know the velocity of light, it is necessary only to measure the distance between the cells at this point.

In practice, this meant squinting through the analyser prism and moving the second cell, which was mounted on a trolley, to a position at which the transient flash from the spark-gap was at its dimmest. This Allison and his colleagues found they could do precisely enough to detect a difference in the time lag of the Faraday effect of 3×10^{-10} seconds (three-hundred million-millionths of a second, or the time that it takes light to travel 10 centimetres). They now began to make measurements of the time lag for a whole series of pure liquids and solutions and quickly the most remarkable results began to emerge and to find their way into the journals. Different substances showed different lag times and many gave rise to more than one maximum, each characterising one grouping in the chemical structure. Allison and his students examined solutions of salts. (These are inorganic compounds that dissociate into their component ions when dissolved in water; for instance, common salt, sodium chloride, in solution is a mixture of equal parts of positively charged sodium and negatively charged chloride ions.) It transpired that different sodium salts produced different lags, the signal coming only from the sodium ions, but modified according as they originated from sodium chloride, sodium sulphate or sodium nitrate. This seemed wholly at odds with established knowledge, but more, the procedure had extraordinary sensitivity and could detect substances at concentrations far below those accessible by other methods of analysis. Yet the signal was not dependent on the actual concentration; it only vanished abruptly below some threshold level.

The pace now became frenetic, for Allison and others found that the method could distinguish isotopes (forms of a chemical element differing in weight—the number of neutrons in the nucleus—but not in chemical properties). Lead salts, Allison found, gave rise in his apparatus to sixteen separate maxima of light intensity, and so lead was a mixture of sixteen isotopes. The result was confirmed in every detail by another laboratory with an improved apparatus. Soon two new elements were discovered and triumphantly named alabamine and virginium. A highly respected chemist, not from Alabama or the backwoods, Wendell Latimer, from the University of California at Berkeley, built his own apparatus and discovered the third hydrogen isotope, tritium, of which there had been some indications from more orthodox directions. Latimer had apparently talked about Allison's exciting new method to Irving Langmuir, who relates in his lecture what happened. Latimer discussed his plans with his colleague at Berkeley, G.N. Lewis, a colossus of physical chemistry, who offered to bet him ten dollars that the whole affair was a delusion. Latimer took the wager and soon presented Lewis with such spectacular results that he was forced to pay up.

Some years later, when the Allison effect was already consigned to oblivion, Langmuir interrogated Latimer about his results. Here is how he recollected Latimer's words:

'You know, I don't know what was wrong with me at that time. After I published that paper I never could repeat the experiments again. I haven't the least idea why. But…those reults were wonderful. I showed them to G.N. Lewis and we both agreed that it was all right. They were clean-cut. I checked them myself every way I knew how to. I don't know what else I could have done, but later on I just couldn't do it again.'

The detection of radioactive decay products and separation of isotopes was hailed as a potential breakthrough for palaeontological dating; organic chemists and biologists were also not to be excluded from the euphoria. Products formed in minuscule amounts in chemical reactions were detected; chlorophyll was found to catalyse new processes, vitamin A was traced in biological fluids and the precise location of ingested uranium compounds in the liver of a rabbit was established.

It was more than five years before serious debate about the Allison effect was joined in the scientific literature, although there had been some murmurs of dissent as early as 1929. Several publications appeared, reporting failure to find any time delay in the Faraday effect. It now emerged that the ground had been covered before with a quite different outcome: fifty years earlier Bichat and Blondlot, no less, who was to come such a cropper over his discovery of the N-rays, had published a meticulous study. They used a polariser, rotating at up to 70 000 r.p.m., and the light of a spark transmitted through a tube of liquid round which a magnetic field was activated by the same discharge, and were able to show that for the liquid in question a time delay, if it existed, was shorter than 10 milliseconds (still a lot more, however, than Allison's apparatus purported to measure). Moreover the Faraday effect was well understood and no theoretical basis existed for lags such as Allison had been observing. More physicists by then were inclining to the adage enunciated by the cosmologist, Sir Arthur Eddington, that no fact should be believed until confirmed by theory.

Now the existence of a measurable time-lag was again questioned, and as time went on more laboratories reported failure to reproduce the positions of the maxima reported by Allison. The settings were dependent on estimates of relative brightness—hazardous, as we have seen—and attempts were therefore made to locate the peak positions of the trolley by intensity measurements with a photocell (which produces a current output in proportion to the light intensity falling on it), and with a photographic plate. There was no correlation, the sceptics announced, between the positions of maxima and minima of light intensity and the nature or concentration of the substance in the tubes. But Allison's followers were not put off. As late as 1936 a series of three papers from no less an institution than Washington University in St Louis appeared in the *Journal of Chemical Education*, which, while conceding that controversy existed, nevertheless proclaimed that in the view of the authors 'a mass of positive data, which has been confirmed by objective means far outweighs numerous negative results whether objectively or subjectively determined'. They further opined that the method was 'perhaps the most important tool for chemical research developed in the last decade...It will detect the presence of a substance in concentrations as low as a few parts in 10^{12}.' To reinforce the point they added a new study of their own, in which a series of solutions of inorganic substances at concentrations down to one part in ten-thousand million were prepared by one experimenter, and designated only by numbers. They were then tested blind by his colleague, who had been given a list of the samples but not the code. He got all of them right and did almost as well in a series of other tests. The probability that the results overall were due to chance was one in 7560.

At the same time, the increasingly numerous reports of failure were met by an argument all too similar to that heard in France during the heyday of N-rays: perceiving the effect required exceptional sensitivity and training. Thus—'A person who plans to begin this type of work should realise that many hours, weeks and even months may be consumed before dependable observations can be made'. Some, it seems, never managed it, while a chosen few succeeded at once. Nevertheless, 'Anyone who enters into this work with the preconceived idea that within a few weeks he will be obtaining positive results, is almost certainly doomed to disappointment.' Gradually, however, reports of new triumphs for the Allison effect became fewer, and it faded like a mirage in the desert, though it was never extinguished by a single, universally accepted refutation. Most physical chemists must have made up their minds that it was a figment of a collective imagination and ceased to worry about it. The *American Chemical Society* announced in 1937 that it would accept no more publications on the Allison effect in its journals.

Langmuir's rules

Irving M. Langmuir (1881–1957) was a remarkable scientist with omnivorous interests, who frequently annoyed his contemporaries by uninvited and brash incursions into their areas of research. He is remembered chiefly for his studies on surfaces, which brought him a Nobel Prize and an eponymous immortality (in an instrument, the

Irving Langmuir

Langmuir trough; a fundamental physical relation, the Langmuir adsorption isotherm; and most recently, a journal of surface science called simply *Langmuir*). Langmuir spent most of his working life at the laboratories of the General Electric Company at Schenectady in New York state, and there he pursued for some years an arcane side-line, the study of what he called pathological science. Langmuir never published anything on the subject and all that remains is the transcript of a lecture that he gave in 1953. An edited version was published in 1989 by one of Langmuir's colleagues at General Electric, and contains, among other things, Langmuir's account (from which I have quoted above) on his encounter with Bergen Davis and Arthur Barnes and his annihilation of their work.

Langmuir identified a series of stigmata by which pathological science could be recognised, and they are these:

1. The maximum effect that is observed is produced by a causative agent of barely detectable intensity, and the magnitude of the effect is substantially independent of the intensity of the cause.

2. The effect is of a magnitude that remains close to the limit of detectability or, many measurements are necessary because of the very low statistical significance of the results.
3. There are claims of great accuracy.
4. Fantastic theories contrary to experience are suggested.
5. Criticisms are met by *ad hoc* excuses thought up on the spur of the moment.
6. The ratio of supporters to critics rises up to somewhere near 50 per cent and then falls gradually to oblivion.

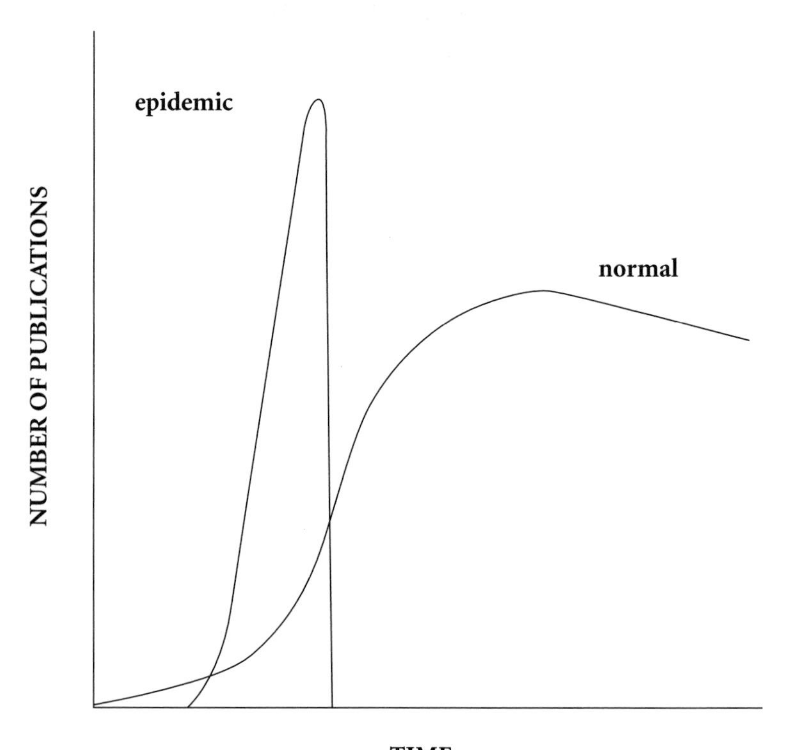

The characteristic time course of interest in a normal discovery and in an illusory (in Langmuir's parlance pathological) discovery, which engenders an epidemic of self-deception. The vertical axis indicates the number of publications on the subject with time (horizontal axis) from the first announcement.

Now it will be clear that all of these criteria apply in every detail to each one of the aberrations discussed so far. Most of them also fit the more recent cases of mass delusion that I shall consider, but we shall also find some examples of large effects with trivial explanations, missed or wilfully disregarded in the urgent quest for a grandiose *coup*. It is also of course difficult to renounce a cherished theory, the product of costly intellectual and emotional investment, and to accept the cost to ambition, reputation and pride of a humiliating retraction. As the economist J.K. Galbraith put it, 'faced with the choice between changing one's mind and proving that there is no need to do so, almost everyone gets busy with the proof'. And so a fatuous optimism triumphs over the caution that must guide all scientists through most of their working lives.

CHAPTER 4

Nor any Drop to Drink: the Tale of Polywater

Water is H_2O, hydrogen two parts, oxygen one,
but there is also a third thing, that makes it water
and nobody knows what that is.

Thus D.H. Lawrence, and when he penned these lines in 1929, physical chemists already in reality had a fair idea of what the third thing was. Water is a highly unusual chemical substance, and, without its curious properties, there would be no life on earth. Water, to begin with, is a liquid at all but the coldest terrestrial temperatures, while hydrogen sulphide, in which the oxygen of water is replaced by sulphur—an element similar to oxygen, but heavier—is a gas. Water freezes at 0°C, a remarkably high temperature, and the solid, ice, is less, rather than more dense than the liquid with which it is in contact. Its specific heat (the heat input required to raise its temperature by a degree) is far higher than would be predicted from the properties of chemically similar substances, and drops by half when the water freezes. The consequence is that as the Gulf Stream flows from a colder to a hotter region it releases energy into the atmosphere. In effect the high specific heat of the water enables it to absorb energy from the sun and so create an energy buffer, which is responsible for the stability of our climate. Without damaging their structure, water will dissolve biological substances, including the giant molecules (polymers), such as proteins, polysaccharides, DNA and RNA; their structures in fact incorporate water, and without it they generally collapse. Water forms itself into a layer, the so-called hydration shell, around these molecules of life, and partly for this reason accounts for some 60 per cent of our bodyweight. (Even the Pope, said J.B.S. Haldane, is 60 per cent water.) Water in our tissues does not turn to ice at 0°C, and cold-blooded Arctic fish can swim in sub-zero water temperatures without fear that their blood will freeze.

All these properties stem from a phenomenon called hydrogen bonding. In the chemical bond between hydrogen and oxygen there is a displacement of negative electronic charge away from the hydrogen and towards the oxygen. This allows the hydrogen to enter into a weak but significant association with the negative charge on the oxygen atom of another water molecule. This can be represented as shown. The consequence is that water does not consist of isolated H_2O molecules, jostling around independently of one another, but rather of a mass of large sticky clusters. When water boils the hydrogen bonds are broken, their energy is liberated and the individual H_2O molecules enter the gaseous state. Hydrogen bonding dominates the properties of water as we perceive them: the hydrogen bond is Lawrence's third thing.

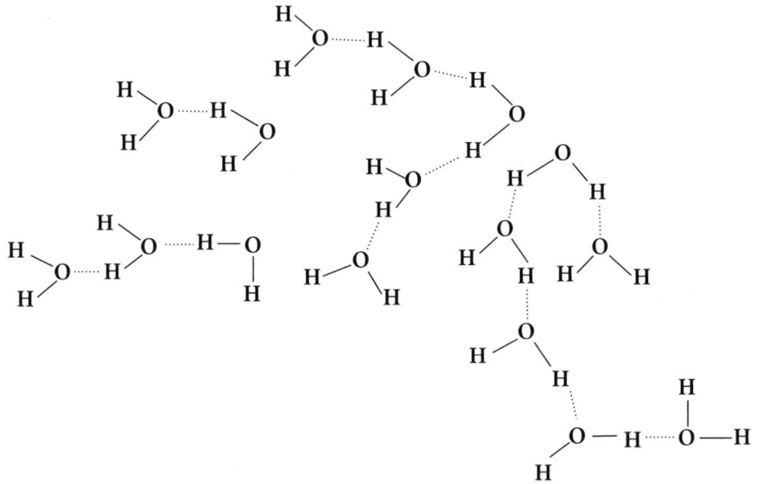

The structure of water: water molecules, H_2O, associate with one another through relatively weak bonds, called hydrogen bonds, between the oxygens and hydrogens of adjacent molecules. Liquid water is thus a mixture of clusters of different sizes ('flickering clusters'). When water evaporates these bonds (dotted lines) are broken and so steam consists only of individual H_2O molecules.

Water was and is the most intensively studied chemical compound and its properties form the substance of innumerable research papers, theses and books. The scientific world was therefore startled by reports that emerged from Russia in the mid-1960s, authored by a highly respected physical chemist, an expert on surface properties, Boris Derjaguin. (For reasons not altogether clear the transliterated French spelling has almost invariably been preferred, although the name also appears as Deryagin.) A form of water had been prepared in his laboratory with properties astonishingly different from those of the water that we know. It was viscous (the viscosity was reported to be fifteen times that of ordinary water), its density was lower by 20 per cent than that of the water we drink, it had greatly increased thermal expansion, it boiled not at 100 but at 250°C, and somewhere below 30°C it gradually solidified, though not to anything resembling ice (which forms sharply at 0°C). The mysterious liquid was referred to as 'anomalous water', or 'water II'.

Derjaguin had been put onto the phenomenon by another Russian, N.N. Fedyakin, working in a technical institute in Kostromo, a remote industrial city east of the Urals. Fedyakin was occupying himself with an old problem—the behaviour of water in very small droplets. The celebrated nineteenth-century physicist Lord Kelvin had observed that water evaporated at an unexpectedly high rate from such spherical droplets, a consequence of the large curvature at the surface. Moreover, the molecules at the surface cannot form as many hydrogen bonds as those in the bulk liquid because they have no neighbours facing outwards, and they also align themselves with their electrostatically charged axes (their electric dipole, as it is called) along the surface. Fedyakin was not in fact looking at water in droplet form, but chose rather to maximise the surface area for a given volume by putting it into very fine glass capillaries. He soon found that water condensed into such capillaries appeared to have increased density. His work came to the notice of Derjaguin, who had a large and internationally known laboratory at the USSR Academy of Sciences in Moscow. Derjaguin resolved to investigate and brought the considerable resources of his institute to bear on the problem. Fedyakin thereupon vanished from view and seems never to have been heard of again (although Derjaguin always gave him full credit for the initial discovery).

Starting from 1962, Derjaguin published a series of reports on the attributes of this strange water, condensed at low vapour pressure in glass or quartz capillaries, less than one-tenth of a millimetre in diameter. (Some, used in later work, had a diameter of only one ten-thousandth of a millimetre.) The accounts appeared in Russian journals and did not at first attract much attention in the West, probably because they gave little indication of what the structure of the 'anomalous water' might be. An investigation of its structure would have demanded the most modern methods of analysis, already relatively commonplace in the West, but not yet widely available in Russia. Nevertheless, a broad range of studies on the physical properties was undertaken. Asked at a conference how much of the material in all had been prepared, Derjaguin replied: 'about enough for fifteen papers'.

Derjaguin's first attempt to tell his story in a Western publication failed, for in 1966 *Nature*, the London-based international journal, rejected his manuscript on the grounds that insufficient care had been taken to eliminate impurities—an unerring criticism, as events were to prove. But then two events forced 'anomalous water' on the attention of scientists in Western Europe and the United States. The first was a lecture tour undertaken by Derjaguin, who by all accounts was a compelling speaker; an hour's verbal presentation is not of course open to the same critical scrutiny as a manuscript submitted for publication, for then the expert reviewers (or referees), selected by the journal, will look for lacunae or evasions and can ask the authors for more information or indeed demand additional experiments before the work is deemed adequate for a place in the scientific literature.

The second and more important factor that compelled a serious examination of Derjaguin's claims was the endorsement of one of the

John Desmond Bernal

most respected figures in the field of molecular physics, John Desmond Bernal. Bernal, half Irish, half Jewish, was a man who attracted adulation and was the subject of endless anecdotes. He appeared in novels, barely disguised (most familiarly as the crystallographer, Constantine, in C.P. Snow's *roman à clef*, *The Search*). Bernal was a polymath of varied and formidable accomplishments, known to his intimates as Sage. He was raffish and exuded a social and sexual allure which brought him a succession of liaisons and a circle of loyal and adoring friends. Most of all, he was a deeply committed communist, eloquent about his beliefs, and in the years before and after the Second World War a political pied piper to many colleagues and especially the young. Bernal was at this time Professor of Physics at

Birkbeck College in London, an institution devoted to serving mainly older, part-time and evening students. There he had built up a strong research group concerned primarily with the study of molecular structure (not least that of biological substances, such as proteins and viruses), and he himself had been preoccupied for several years with the structure of water.

Bernal wrote highly flattering accounts of the virtues of science in the Soviet Union, and he attracted some opprobrium for his failure to condemn the ravages of Lysenko (Chapter 9). It was clear that he strongly desired the success of Soviet science, and when Derjaguin came up with what appeared to be a stunning development in his own special area of interest, Bernal's critical judgement went by the board. Derjaguin's anomalous water, Bernal declared, was 'the most important physical chemistry discovery of the century'.

Derjaguin had in fact made some careful and ingenious physical measurements of such properties as the index of refraction (startlingly high) of his 'water II', as he preferred to call it, in fine capillaries. He also measured its molecular weight, which came out at about 180 (whereas H_2O works out at 18—sixteen from the oxygen atom and one each from the hydrogens), implying that about ten water molecules were strongly linked to one another. Derjaguin was sensitive to the criticism that the substance he was studying was no more than water heavily contaminated with impurities. He could exclude this, he asserted, for several reasons: first, simply sucking water into capillaries (rather than condensing it from the vapour) generated no anomalous properties, even when it had been kept for hours under pressure at 400°C, nor did contact with powdered glass. The second argument was that when the anomalous water was vapourised and heated to 900°C, it reverted when condensed to ordinary water. Furthermore, extraneous impurities had been excluded by doing the preparations in a glass-blown apparatus with no greasy stop-cocks, and placing two liquid-nitrogen traps (to condense any volatile impurities) between the apparatus and the vacuum pump used to lower the vapour pressure. He had also measured the surface tension and found it higher than was compatible with dissolved organic impurities.

The bandwagon now began to roll, as physicists and chemists in the West jumped aboard. A strong stimulus came from a paper by a respected American research group, reporting that the infrared spectrum of the material was quite different from that of water.[1] One American researcher realised what was afoot when he discovered that the Fisher Scientific Corporation, the largest supplier of laboratory glassware, was found to have run out of desiccators—vessels that can be closed to the outer atmosphere and evacuated, and are most often used for drying chemicals, or, as then, to create an environment of controlled vapour pressure. There were loud avowals of success in reproducing Derjaguin's phenomena, and studies by spectroscopic and other techniques were reported to show that the anomalous water consisted of molecules stably associated into large clusters. The substance became known henceforth as *polywater*. Its allure was perhaps greatest for the thoreticians, who now got busy calculating what its structure might be and why it should be stable, rather than reverting to ordinary water. Inevitably polywater was found to occur in living cells, or at least in association with enzymes. It was discovered that anomalous water had been described by a physicist from Howard University in Washington, D.C. in 1928, and the Bangham brothers, both well-known scientists in Britain, recalled that their father had made related observations in 1937, while Professor of Physical Chemistry at the University of Cairo; he had directed a jet of superheated steam at a clean sheet of mica and had observed anomalous properties in the thin layer of water that condensed on the surface. The discovery, evidently, was ahead of its time and the seed had fallen on stony ground.

Not all scientists were as well disposed as Bernal to the Russians, and the Cold War, it should be recalled, was then at its height. Several laboratories sought to advance their own claims and disparage the Russian work that had inspired their efforts. The press release was enlisted as an instrument of self-advertisement. The *Miami Herald* ran a headline, announcing 'Miami Scientific Team Creates Mysterious New Form of Water'. The popular press worked itself up into a state of high excitement. The *New York Times* indulged in a particularly extravagant flight of fancy: 'A few years from now', it enthused, 'living room furniture may be made out of water. The antifreeze in cars may be water. And overcoats may be rainproofed with water.' A Russian astrophysicist conceived the idea that luminous ('noctilucent') clouds, high in the atmosphere, were made of polywater droplets, which is why they remained there without either evaporating or crystallising. And one of his American *confrères*, F.J. Donohoe, conjectured that the water on Venus might all be in the form of polywater.

There were also prognostications of doom. Not long before, Kurt Vonnegut had published to critical acclaim his scientific novel *Cat's Cradle*, in which occurs a new allotrope[2] of ice, ice IX. This does not melt at 0°C, and liquid water to which a few drops have been added turns at once to ice IX and crystallises; so a little ice IX, tipped into the sea, would cause the oceans to solidify and liquid water, and therefore life, to vanish from the earth. The parallel between polywater and Vonnegut's flight of imagination did not escape the notice of journalists, or evidently of some scientists: F.J. Donahoe, writing in *Nature*, made the stark declaration that 'I regard this polywater as the most dangerous substance on earth'.

Most scientists were less agitated. Joel Hildebrand, the Nestor of the American scientific establishment (who was to write his last paper at the age of 100) could not take seriously a transformation in water caused by contact with glass; ' We choke on the explanation', he wrote in the journal *Science*,

> that glass can catalyse water into a more stable phase. Water and silica [the substance of sand and minerals, such as quartz] have been in intimate contact in vast amounts for millions of years; if a more stable kind of water were possible, it is hard to understand why any ordinary water should exist...It is easy to see why a spectroscopist might be excited by the term 'polywater' to try to design ways for water to

> polymerise which nature had overlooked, but I think a chemist who feels curious about what is in those glass capillaries would have more success if he assumes that he is dealing with a system of two components

—by which he meant water and silica. Some people were troubled by the apparent violation of the Second Law of Thermodynamics (the impossibility of spontaneous conversion of a system from a state of lower to one of higher energy), which they saw in the formation of polywater—a form presumably less energetically favoured—in equilibrium with water, since the capillaries were open and received water vapour from ordinary water. The most famous physicist in the land, Richard Feynman, had his own comment on the energetics of the system: 'There is no such thing as polywater because if there were, there would also be an animal which didn't need to eat food. It would just drink water and excrete polywater.'

All the same, the United States Office of Saline Waters found it prudent to devote 10 per cent of its considerable budget to promoting research on polywater. The Director later confessed that he too had had his doubts; but how, he asked, would he have been judged if the discovery had turned out to be real and his agency had not pursued it? Here, as in so many of these examples of seemingly reckless credulity, the Director is articulating the ineluctable principle of Pascal's Wager. Pascal regarded atheism as a folly, for however minute the probability that a vengeful Christian God existed, the consequences of disobeying his dictates, if he did exist, would be so immeasurably great that submission to religion was the only rational course. And similarly, the greater the reward of winning, the longer the odds that the gambler is willing to accept.

More than 500 papers were published on polywater and some 400 scientists in greater or lesser degree staked their reputations on it before the bubble burst. The theoreticians suffered worst, for they, as a class, are congenitally bored by experimental detail and so accepted too readily the claims of insufficiently critical experimentalists. Experimental scientists, on the other hand, tend to believe that their

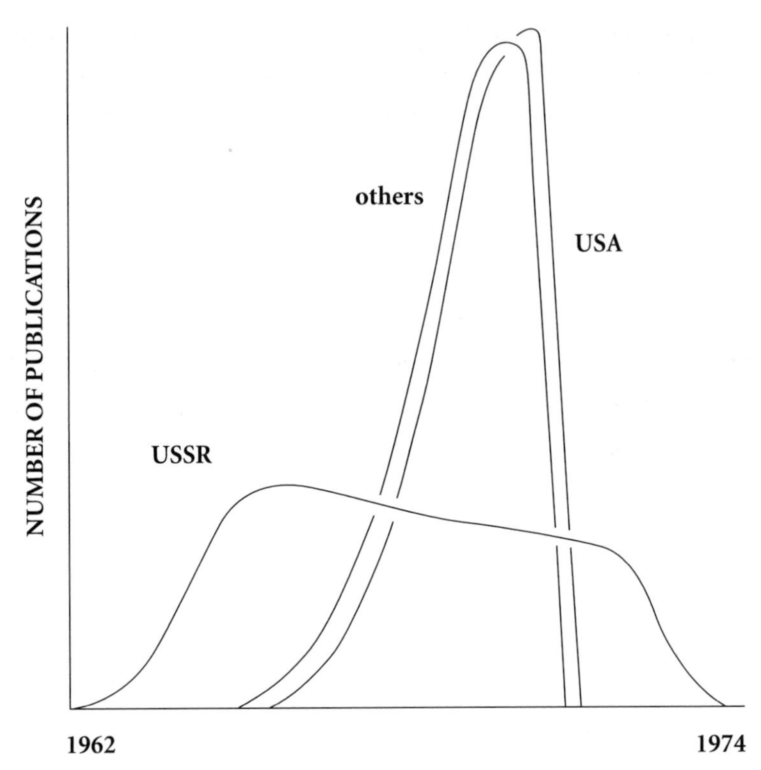

Time course of the polywater epidemic: the graphs show the number of papers published on polywater with the passage of time after the first announcement of the phenomenon. Initially all reports were from the Soviet Union and were largely ignored in the West. There was then a sudden upsurge of interest in the United States and other Western countries, leading to an effusion of publications that far surpassed the number being published in the Soviet Union. When the polywater phenomenon was shown to be a figment due to impurities, the work in the West abruptly ceased amid embarrassment and recriminations, but it took some further time for the Russians to concede defeat.

theoretical colleagues can conjure up an explanation for anything, and could on demand produce the opposite conclusion with equal conviction. Nevertheless, polywater presented what seemed to be a challenge of exceptional interest, and it engaged the attention of some of the most eminent practitioners in theoretical chemistry. Writing during the period of agonising introspection that followed the final evaporation of polywater, the distinguished theoretician Leland C. Allen told of the excitement that the appearance of something so radically new and unexpected had engendered. The attention that his first paper on the subject had attracted 'gave me a more convincing sense of having made a true contribution to science progress than I have felt for some other conventionally successful research projects'. He had been aware that many experts, such as Joel Hildebrand, had scoffed at polywater. 'The sceptics', Allen commented, 'were accusing the believers of naïveté and gullibility, and the believers were accusing the sceptics of arrogance and lack of imagination.'

By the early 1970s spectroscopic analyses of polywater preparations were starting to appear in the journals, showing the presence of silica (derived from glass) and other impurities. As early as 1970 the US Naval Research Laboratory had voiced doubts about the authenticity of polywater. Derjaguin reacted with indignation: he would not, he wrote, have wasted nine years working on an artefact. He could discount any suggestion that polywater was a suspension of silica on the basis of direct observations on expansion of water trapped between silica particles. Finally, Dennis Rousseau from the Bell Telephone Company's laboratory in New Jersey published a devastating report, the burden of which was that polywater samples contained anywhere between 20 and 60 per cent by weight of foreign substances. These included the common ions (sodium, potassium, chloride, sulphate and so on), in about the proportions found in sweat. The spectra of polywater in fact closely resembled those of sweat. (Rousseau acknowledged generous donations of sweat samples from colleagues.)

The situation was in reality a little more complex, for the champions of polywater in America had found that the substance could equally well be produced in glass or in quartz capillaries, and Rousseau's experiments had been performed on preparations made in glass. Several other research groups had already suggested in 1970 and 1971 that such samples were full of salts; sodium acetate, for instance, was identified as one abundant impurity, formed perhaps on washing the glass with ethyl alcohol. This might be oxidised in the capillary to acetic acid, which would leach sodium out of the glass. Sodium acetate might also have been produced by the action of an oxyacetylene flame on the glass, when it was melted to draw the capillaries. A pair of researchers in Australia actually suggested at a quite late stage that there were different kinds of polywater, in particular 'glass polywater' and 'quartz polywater'. In some sense this was evidently true. Derjaguin himself had chosen quartz for its much greater chemical stability and had gone to great lengths to exclude ionic (see p. 75) contamination. But quartz evidently presented different problems, physical rather than chemical: the very high surface area of quartz in contact with water in the tiny capillaries allowed minutely particulate material to be shed from the walls. Now 'anomalous water' was held to be a mixture of ordinary water and polywater (a term Derjaguin never espoused, for it carried the presumption of long strings or stable clusters of linked water molecules, for which there was little evidence). When the water was pumped off, a viscous or even gelatinous residue—pure polywater—remained. When this was examined by more rigorous means than the zealots thought it necessary to employ, it was found to have all the properties of a silica gel, a kind of colloidal (see p. 39) vaseline-like jelly, normally produced from finely divided silica (chemically identical to quartz) and water. Polywater could then, it appears, be generated with more than one kind of impurity. With hindsight this seems not so very surprising, because there are few materials from which soluble substances cannot be leached out or particulate matter thrown off. In the capillaries a lot more of the water is in contact with the glass or quartz than in a large vessel. For example, the wall area surrounding one millilitre of water in a test tube of one centimetre diameter is about four square-centimetres, whereas in a typical capillary, used for preparing polywater, the corresponding wall area might be a thousand times as great—and the total capillary length required to contain the millilitre would be perhaps 500 metres; and in the smallest capillaries these numbers would be hugely higher. So, a sizeable fraction of the water would be in contact with the glass or quartz and in this surface layer a solution or suspension would form. As more water condensed from the vapour above, this dense fluid would sink and form a visible lump in the tube, exactly as the experimenters observed.

In 1973 Derjaguin and his colleage, Churaev, in a short note in *Nature*, honourably conceded: the properties of polywater, they had found, could be explained by a contaminant of colloidal silica. Not everyone accepted the explanation and, one letter to *Nature* pointed out, the grossly anomalous properties reported by Derjaguin and his associates in a long series of papers could not possibly be accounted for by the amounts of silica they now confessed were present. The explanation was not chemical but psychological: it was a case of self-deception, just like the N-ray affair. On the whole, however, the scientific community (except in the Soviet Union) treated Derjaguin with understanding; not long ago a special symposium on surface science was held in his honour in celebration of his ninetieth birthday.

Notes

1 The infrared spectrum—the wavelengths of light in the infrared region of the spectrum (the wavelength range above that of visible light) absorbed by a substance—is related to the frequencies of oscillation of the various chemical bonds in the molecule and reflects its chemical structure.

2 Allotropes are crystalline forms of the same substance, each characterised by a different arrangement of the molecules relative to each other.

CHAPTER 5

The Wilder Shores of Credulity

High intelligence and learning do not always exclude unreason. From its earliest beginnings science was recognised to depend on scepticism. Reports of phenomena that appear improbable within the prevailing paradigm, as Thomas Kuhn termed the current viewpoint in a field of science, need to be tempered in the furnace of doubt, even hostility. When this is overcome, what Kuhn calls a paradigm shift ensues. But if the prevailing paradigm is too willingly overthrown, anarchy results and progress becomes impossible. The conscientious scientist adopts in fact the criterion suggested by David Hume for judging miracles: is it more probable that a miracle has occurred than that you have been deceived? So in eighteenth-century France, when Franz-Anton Messmer persuaded a gullible public of the existence of animal magnetism and harnessed it to cure diseases, King Louis XVI ordered a Royal Commission to be set up to study and report on the matter.

Messmer did not invent the theory of animal magnetism: he learned it from a turbulent Austrian priest, Father Hell, though it probably originated with Paracelsus in Germany. Messmer's assertion, though, that he could transmit his own superabundance of magnetic fluid to patients on the couch and thereby cure them of a host of maladies, and that he could by the power of will magnetise paper, wool, bread and animals, not to say charge up Leyden jars, brought him only ridicule in his native Vienna. But in Paris he secured the patronage of a fashionable doctor, who helped him to establish a salon, decked out with extravagant oriental furnishings, and to attract rich patients. Messmer's theatrical ways began to draw crowds and often occasioned hysteria and life-giving convulsions in his patients. Imitators sprang up around France and elsewhere in Europe, some of them confidence tricksters, others, including two woolly-minded French aristocrats, self-deluded. In most major cities so-called 'Societies of Harmony' were founded, dedicated to therapy by magnetism.

Impassioned debates ensued, and the Faculty of Medicine of the Sorbonne finally denounced Messmer as a charlatan. His patron, Dr d'Eslon, thereupon urged him to challenge the Faculty to a contest. The Faculty agreed and proposed that a group of twenty-four patients be chosen, twelve to be treated by Messmer and twelve by its own members. The Government threw in the offer of a pension and an order of chivalry if Messmer could give demonstration of the efficacy of his treatment and make his secrets openly available. Messmer prudently rejected the offer and abruptly left Paris. Dr d'Eslon, however, did not repine; though threatened with expulsion from the Faculty, he insisted that Messmer's discovery was of momentous import and called for a formal investigation into his claims. The King reacted by ordering a Royal Commission of the Faculty of Medicine to be assembled, and then a second by the Faculty of Science. The latter nominated a group of eminent *savants* which included Lavoisier (all too soon to lose his head to the guillotine), Bailly and Benjamin Franklin. Messmer had no option but to return from his retreat in Spa. The *savants* observed, experimented and took evidence for five months, at the end of which they gave their verdict: '*L'imagination fait tout, le magnétisme nul*'. Messmer's reputation could not survive this unequivocal judgement and he departed hastily for Austria, never to return.

This of course did not stop the magnetisers, who reached England a few years later. The first seems to have been an opportunistic disciple of Messmer, Dr Mainauduc, who lectured to packed houses and founded the Hygeian Society. Later, towards the beginning of the nineteenth century, an American doctor, Benjamin Douglas Perkins, enjoyed a vogue in England. He treated his patients with a patent 'metallic tractor', with which he directed the passage of magnetic fluid. His career was extinguished by an English doctor, who demonstrated that he could make his patients better by using a wooden replica of Perkins's instrument. His pamphlet, 'Of the Imagination as a Cause and Cure of Disorders, exemplified by Fictitious Traction', may have been the first explicit description of the placebo effect (p. 154).

In France, animal magnetism enjoyed a revival during the second decade of the nineteenth century. An evidently well-meaning manual for practitioners aroused wide interest. It urged that the way to success was to shun all doubts and to concentrate on a favourable outcome. But it was especially important to ensure that no troublesome or inquisitive witnesses be present, nor should patients with such destructive attributes be treated. This injunction uncannily prefigures investigations of so-called paranormal phenomena in our time (p. 103).

Early in the nineteenth century the magnetisers were for the most part supplanted by 'electricians', as they were called. They drew their inspiration from the experiments of Galvani and Volta on the twitching of excised muscles (frogs' legs, for example) in response to an electrical impulse. It was these observations that also provoked attempts to revive the recently dead by similar means. When the cadavers twitched it was regarded as a sign of incipiently returning life. A literary outcome of these dramatic manifestations, and one that was to colour the public perception of science and its votaries for the next century, was Mary Shelley's *Frankenstein*. Its publication heightened the general interest in the life-enhancing attributes of electricity, and one of the most prominent beneficiaries of the public's eager desire for a dramatic revelation was Andrew Crosse. He was a gentleman-amateur scientist, born into an intellectual family of radical inclinations—his father had been a friend of two great radical scientists, Joseph Priestley and Benjamin Franklin—and he set up a laboratory in his ancestral seat in the West Country. Crosse's circle of friends included Sir Humphry Davy, who may have influenced his scientific endeavours. Crosse's interest was in electricity, and in particular he conceived the theory, which he set out to test, that crystalline mineral deposits were generated by electric discharges. He passed a current from a voltaic cell through various solutions of inorganic substances, especially silicates, in the hope of forming quartz crystals at an electrode.

Crosse thought he had succeeded and in 1836 a report of his work to the British Association (the main annual forum for the presentation of scientific discoveries in Britain) was introduced by the Professor of Geology at Oxford, William Buckland. This was a time of great ferment in geology and Crosse's observations were thought to throw light on the mechanisms of formation of mineral veins and stratification, at the time a hotly debated subject. His work was commended by Buckland and the doyen of British geologists, Adam Sedgwick.

That same year Crosse made another discovery. After a day of electrolysis of an inorganic solution, inspection of the electrodes revealed the formation of some 'whitish excrescences or nipples'; after four days these were enlarged and had thrown out filaments; at twenty-six days 'each figure assumed the form of a perfect insect' and another two days on, the insects came to life and scuttled away. Other kinds of electrolysis experiments, with different electrodes and electrolytes, also generated the selfsame insects. Crosse must have realised that his results would elicit incredulity and probably ridicule and he did not publish them, but he told his friends and word leaked out. The journals of the day, such as *The Times*, the *Gentlemen's Magazine* and the *Athenaeum*, seized on the story, and Crosse was soon forced to give a full account. Whether life forms could arise by spontaneous generation or only from a living precursor (*omnis cellula a cellula*, as the tag had it) was then still a matter of fierce dispute, and Crosse's critics recognised that as evidence for spontaneous generation his experiments were quite inadequate. He had not, for instance, excluded air, and the insects could, in any event, have hatched from eggs that had survived the electrolysis process. Faraday was reported in the press as being convinced by the evidence, but he denied this angrily in a letter to *The Times*. Attempts to repeat Crosse's experiments followed, and, as always in such situations, several enthusiasts reported success. Eventually, in 1837 at the British Association meeting, John Edward Gray, a leading naturalist of the day, and others, denounced Crosse's evidence as worthless, and although some debate ensued, this should have been an end of the matter.

Entomologists, however, were still curious to know what manner of insects had crawled out of the electrolysis bath, and Buckland collected a consignment of the creatures from Crosse and passed them to the celebrated anatomist and naturalist (who was to become Darwin's most tenacious opponent) Richard Owen. Owen looked down his microscope and saw what he recognised as cheese mites. A botanist at the Academy of Sciences in Paris thought otherwise: this was a previously unknown species of the same genus as the common cheese mite, and was christened *Acarus horridus*. While naturalists for the most part remained implacable, the medical profession was drawn to the doctrine of spontaneous generation, held to be responsible for the appearance of parasites, such as tapeworms. A surgeon, William Henry Weekes, threw himself into a programme of experimentation to generate insects under better controlled conditions than Crosse's. His electrolysis cell was confined in a closed container over mercury. In 1841 he announced that, after prolonged electrolysis of a potassium silicate solution, five fine living insects had been recovered from the apparatus, together with a mass of quartz crystals. Crosse himself performed similar experiments with like results. The outcome again caused a stir in the press, and as late as 1850 an article in the leading medical journal, *The Lancet*, judged that the evidence for spontaneous generation of insects was now incontrovertible.

Adam Sedgwick, who had at first supported Crosse, now took fright, and urged that these impious attempts to dabble in an area—the origin of life—beyond the permissible limits of human curiosity be forthwith discontinued. (In this he was at one with the populace of Crosse's village in Somerset, who denounced him as a Frankenstein, a 'disturber of the peace of families' and 'a reviler of our holy religion'.) Other geologists thought differently, and the pot was stirred by a report from a German naturalist that siliceous rocks were made up (like limestone) of small dead creatures, in this case insects. Perhaps then electricity could indeed, as many of the 'electricians' believed, revivify dead life forms? Michael Faraday was scandalised by such nonsense, but several reputable scientists felt the need to satisfy themselves by setting up their own experiments. Gradually truth prevailed, a closer inspection of siliceous rocks revealed no insects, and no more was heard of the cheese mites (although it was by no means the end of the spontaneous generation debate: this was settled for practical purposes by Pasteur in 1864, much applauded by Faraday. A few diehards held out, prominent among them a biology professor at London University, E.W. MacBride, who was silenced only by death in 1943).

In London in 1837 a group of aficionados had formed the London Electrical Society, which held meetings and published its proceedings. Crosse was a member, together with a number of reputable scientists, but Faraday never joined. Crosse and his friends believed that 'electrified water' possessed remarkable invigorating and curative properties. It could, for instance, ward off typhus, it was wholesome to drink and, Crosse believed, there might be much virtue in bathing in it.

In the later part of the nineteenth century a new illusion gained a grip on the public imagination: spiritualism took in untold numbers of intelligent laymen, and not a few distinguished scientists fell prey to the wiles of fraudulent mediums. Prominent among these were Charles Richet (a physiologist who was to receive a Nobel Prize for his discoveries in immunology) in France, the German pioneers in the psychology of visual perception, Weber and Fechner—the latter by then almost blind, yet willing to believe that the notorious medium, Henry Slade, was making a piece of string tie itself into knots—Sir William Crookes (who gave his name to the Crookes tube and made enduring contributions to chemistry and to physics), and another leading physicist, Sir Oliver Lodge. In Lodge's case there were mitigating circumstances, for his interest in the subject became obsessive only when his son, Raymond, fell in the First World War. His communications through a medium with Raymond and the revelations that emerged about life on the other side (which resembled a sojourn in a comfortable gentlemen's club) attracted much ridicule.

Interest in the occult, or, as it became known in more pretentious circles, the paranormal, did not diminish with the exposure of the fêted mediums of the *haut monde* as particularly blatant charlatans. It continues yet, and several laboratories still pursue the phenomena with pseudoscientific zeal. When investigated, these studies have invariably been found wanting in objectivity and rigour. Such work has nearly always gone on in isolated locales and has seldom attracted the notice of any significant number of scientists—although even reputable centres of learning have been known to harbour professors with a belief in telekinesis or alien abductions. Attempts to study telepathy and 'distant reading' with proper statistical evaluation of the experimental results persisted for many years at Duke University in North Carolina, where J.B. Rhine held court. Rhine was by general consent honest, but his analytical procedures did not stand up to professional scrutiny, and his voluminous published data, tainted by accusations of fraud against some of his associates, have long since been consigned to oblivion. Effusions of the same kind still emerge from the Stanford Research Institute (no relation to Stanford University), where two engineers, Harold Puthoff and Russell Targ, remain doggedly attached to the problem. Highly entertaining exposés of psychical research from Rhine onwards can be found in books (p. 315) by Martin Gardner (who founded a journal, *The Skeptical Inquirer*, dedicated mainly to analysing such claims) and the magician James (the Amazing) Randi. Both are indispensable reading for anyone interested in the odder infirmities of the human psyche.

Psychic research in sum seemed, with the death of Rhine, to have finally vanished into the refuse-heap of science. Yet in the 1970s there was an embarrassing outbreak of reports from reputable academic sources of paranormal phenomena. In Russia two mediums, Kulagina and Kulashova, performed blindfold reading tricks and pulled matchboxes on fine threads to such effect that their claims to paranormal powers were endorsed by the Soviet Academy of Sciences. A wider outbreak of fatuous credulity occurred in Britain. The individual participants in this absurdity were almost certainly driven to their excesses by the knowledge that they were not alone.

The episode probably had its origins in the appearance on television of a personable Israeli illusionist and former paratrooper, Uri Geller. He was a consummate performer with the ability, possessed by all the best magicians, of distracting the eye with the activity of one hand, while the other performed the trick. His specialities were extrasensory perception, or ESP—receiving messages, for instance, about what was drawn on a card—and causing metallic objects to deform mysteriously. The main difference between Geller and other magicians, however, was (and remains) his claim to true paranormal powers: his magic is not an illusion, it is real, and he has taken his critics to court for saying otherwise. In the wake of Geller's performances there appeared a horde of amateurs and impostors, mainly children, who gave out that they too had the gift of bending metal by feeling it and focusing their energy on the task, not to mention, often enough, the ability to re-start defunct watches.

What followed was high comedy, culminating in the collapse of stout and pompous academic parties and the tinkle of splintering reputations. In the mid-1970s several books on the paranormal appeared. Among them were Geller's own account of his accomplishments, another about Geller's supernatural abilities by his patron, a writer on psychic matters, Andrija Puharich, a counterblast by James Randi (which got him embroiled in an expensive court case) and, extraordinarily, *Superminds* by John Taylor, Professor of Mathematics in the University of London. Taylor had allowed himself to believe that Geller, and more especially a large number of supposedly guileless eleven-year-olds, could bend keys and cutlery by thought alone. Taylor had several accomplices in his study of such phenomena, notably the highly respected theoretical physicist, David Bohm, and another Professor of Physics in London, J.B. Hasted (who believed, amongst other things, that concentrated thought had managed to slow down a clock in his department). In a report in *Nature* they and several colleagues opined that the case for Geller's magic was inconclusive, for the observations had not been carefully enough controlled. A more critical approach was needed than had been exercised hitherto. But, they informed their readers, it had been borne in on them that the accepted methods of scientific investigation did not suffice in this area of research: since the effects that they were seeking to analyse, if indeed they existed, were exhalations of the mind, a certain empathy with the performers was needed. The 'psychokinetic phenomena cannot in general be produced unless all who participate are in a relaxed state. A state of tension, fear, hostility, on the part of any of those present generally communicates itself to the whole group. The entire process goes most easily when all those present actively want things to work well.' More, 'the attempts to concentrate strongly in order to obtain the desired result (the bending of a piece of metal, for example) tends to interfere with the relaxed state of mind needed to produce such phenomena. It appears that what is actually done is mainly a function of the unconscious mind, and that once the intention to do something has been firmly established, the conscious functions of the mind, in so far as they have some bearing on the goal, tend to become more of a hindrance than a help.' The writers here seem to accept from the outset the existence of the phenomena that they propose to investigate: if they do not appear, it is the fault of the observer. The language is not that of scepticism, which is forbidden by the new rules of the investigative game.

In their account Hasted and his colleagues reluctantly recognise that, at least when confronted with a conjurer of Geller's calibre, they could be fooled. Experiments must be designed then to eliminate any such hazard, and they tried this with Geller:

> In the first stages of our work we did in fact present Mr Geller with several such arrangements, but these proved to be aesthetically

> unappealing to him. From our early failures we learned that Mr Geller worked best when presented with many possible objects, all together on a metal surface [We may guess why, but not so the learned authors]; at least one of these objects might appeal to him sufficiently to stimulate his energies. In a later session, we had such a set-up, which included two small plastic capsules, each containing a thin disk of vanadium carbide single crystal. A clearly observable change in the disk within one of the capsules was brought about when Mr Geller held his hands near them...we conclude that this was something no magician could have done.

But then they go on to concede that they are no conjuring experts, 'so that if there should be an intention to deceive, we may be as readily fooled as any person'. This modest understatement is followed in the article by a rejection of the suggestion made by Randi and others that a competent magician should be present throughout all such investigations, because 'A particular magician could therefore say at most that he knew of no tricks that could have brought about a given set of observed phenomena'. The possibility that the magician *could* enlighten them about the nature of the trick that had been played does not enter into their thinking. Moreover, they have discovered that 'magicians are often hostile to the whole purpose of this sort of investigation, so that they tend to bring about an atmosphere of tension in which little or nothing can be done'. The paranormal phenomena, they are saying, only occur before those who believe that they are genuine. The same principle, as we have seen, held true for the perception of N-rays and mitogenic radiation.

John Taylor voiced similar opinions, and urged his readers not to worry excessively about the problem of reproducibility, too often, he thought, seen as a hallmark of quantitative experimental science. Mentally driven processes were simply different from the science to which one was accustomed, and the subjects had to be treated with patience and sympathy. Bright light was not conducive to the right state of mind, nor must the subjects be closely watched, on account of the deadly 'shyness effect'. These innocent eleven-year-old children, unaware of their exceptional gifts, could ravage a canteen of cutlery by the power of the mind, but not while they were being observed. Taylor stated that conditions could be devised to exclude any possibility of fraud. All the same, he was content to leave the children in a room alone with the spoons, or even allow them to take the cutlery home. To render fraud altogether impossible Taylor enclosed some metal objects in a tube.

When James Randi—who had already persuaded audiences that his own demonstrations of spoon-bending were proof of paranormal powers—visited Taylor, he found that the stopper fell out of the sealed tube even before he got to grips with it. Taylor reacted indignantly to Randi's strictures on his procedures. What, he pompously asked in a letter to *Nature*, were Randi's scientific qualifications? He quoted Sir William Crookes's adage that one must trust one's senses, and 'I maintain that a physical enquirer is more than a match for a professional conjurer'. A correspondent asked whether Crookes's pronouncement came before or after he had 'disgraced the scientific calling by becoming an accomplice of the false medium Eliza Cook; if the latter, was it before or after the conjurer Maskelyne indicated the mechanism of deceit?'

But in any event, nemesis was at hand: two workers at the University of Bath, Pamplin and Collins, had tested six of the praeternaturally empowered children in the laboratory. The room was equipped with one-way mirrors, through which the activities of the children could be observed and filmed. With observers in the room the children stroked the spoon 'in the approved manner' between finger and thumb, and reported that the metal was becoming soft, or felt like plaster or water. When the observers gave the appearance of relaxing their vigilance to induce a favourable milieu for paranormal activity, the children went to work. The cameras recorded how one child placed the spoon under her foot and bent it, while others used both hands. Only one of the children did not resort to cheating, and he, alas, bent no metal.

John Taylor in his response implied that what Pamplin and Collins had done was not cricket. Most children, he now averred, cheated in any activity. Only one in six of the children who claimed metal-bending abilities were found to have spoken the truth when tested. (The others had cheated.) Some of the Bath group of children claimed that they had the power, even after being caught out. So perhaps they really did. This curious argument had been deployed by (among others) students of spiritualism, who had even invented a special term, 'mixed mediumship', for the activities of mediums caught red-handed. The procedures followed by Collins and Pamplin were also, Taylor asserted, inadequate. In any case, rigorous tests must be carried out on the mutilated cutlery to establish whether structural changes had occurred in the metal, and the Science Research Council of Great Britain had refused to fund such studies. How then to gain access to expensive equipment, such as an electron microscope? A public offer was made in the form of a letter to *Nature* by Dr Ball of the Department of Metallurgy and Materials Science at Imperial College, but if it was taken up nothing was ever heard of the outcome.

As late as 1978–9, however, *Nature* allowed Taylor and his colleague, E. Balanovski, space to expound on the energetics of paranormal phenomena. Taylor had decided that the only basis for any such effects was electromagnetic radiation of low frequency (microwaves, radiofrequency waves): his subjects would have to harbour a transmitter in their brains, the output of which he decided he could pick up in the laboratory. The emission would increase while ESP was in progress, or during telekinesis or spoon-bending or any other such unexplained manifestation. Since nobody has ever produced a convincing demonstration of any of these phenomena, the experimenters would not have known whether anything paranormal was taking place (unless of course they should have chanced upon it definitively for the first time ever). Therefore it was uncertain what the failure, as duly reported, to detect any emission of electromagnetic radiation might have meant. In one type of experiment the subject was invited to make a suspended needle in a jar rotate. The needle, when watched closely, began to move, but this was traced to static and was stopped by smearing anti-static gel on the vessel. A piece of straw could be made to rotate on the surface in a beaker of water, but this too ceased when an electric fire warming the room was switched off. Metal-bending occurred only under 'relaxed conditions' (when the subject was not being watched), and Taylor also recruited a 'psychic healer' and monitored his emissions when he was on the job. But nobody emitted any detectable radiation at any stage.

The final utterance from Taylor, again with Balanovski, was a theoretical consideration of what energies might be involved in paranormal phenomena. He first dismisses metal-bending as being far beyond the energy range of known radiations, and he goes on to demolish the rationale for all other paranormal effects. It might be asked why Taylor had not talked himself out of the whole profitless venture closer to the outset, but at least in this last publication he abruptly distances himself from his past with this lofty declaration: 'there is no reason to support the common claim that there still may be some scientific explanation which has as yet to be discovered. The successful reductionist approach of science rules out such a possibility except by utilisation of energies impossible to be available to the human body by a factor of billions'. And then the revolving door disgorges us on the same side at which we entered six years or so earlier: 'We can only conclude that the existence of any of the psychic phenomena we have considered is very doubtful'. R.I.P.—for the moment at least. Twenty years might be a reasonable estimate for the time that will elapse before the next reports of paranormal happenings in the physics department of some grave seat of learning.

Attempts by the protagonists to justify all the futile activity, and salvage some of their intellectual credibility, on the grounds that the phenomena needed to be seriously investigated before they could conscientiously be discounted, carry little conviction. To resurrect the paranormal as a fit subject for scientific research on the credit of a conjurer and a gang of eleven-year-olds with the mischievous sense of humour of their kind betrays an unfathomable credulity: as well investigate the possibility that the magician who pulls yards of coloured streamers out of his sleeve grows silk very rapidly in his armpits.

In 1983 British universities competed for the privilege of founding the first professorial chair in the paranormal under the will of Arthur Koestler, the pan-European intellectual and polymath with a penchant for the occult (as his ludicrous book, *The Roots of Coincidence* too plainly revealed). Edinburgh University won the contest. A professor of electrical engineering in the City University in London made a pronouncement in his bid, on behalf of his institution, to the effect that paranormal phenomena have often been demonstrated, but are not yet reproducible because the parameters [sic] for the effects are not known. 'It's like electricity in the days before Faraday.' If you believe that…

CHAPTER 6

Energy Unlimited

The most recent and globally spectacular outbreak of self-delusion—the triumph of desire over reason—began some ten years ago with the news that fusion of hydrogen had been achieved in something not much more complex than a test-tube. Atomic fusion is the process that generates the heat of the sun and the energy released by the explosion of a hydrogen bomb. If it could be brought about cheaply and at a controlled rate, the world's energy needs, no matter how extravagant, would be satisfied in perpetuity, the scourge of pollution conquered and the man who discovers how it is done would be adulated by generations to come as mankind's greatest benefactor. The rewards are ample and the prospect sufficient, as the cold-fusion aberration showed, to unhinge the keenest intellect.

The principal *dramatis personae* were two physical chemists, Martin Fleischmann and Stanley Pons, and a physicist, Steven Jones, and the first act of the drama unfolded in the state of Utah. Martin Fleischmann, born in Czechoslovakia, is British and was Professor of Chemistry at the University of Southampton. His speciality was electrochemistry—the study of chemical and physical processes at electrodes and of ions (see p. 75) in solutions. He was by this time in his sixties and had achieved international eminence in his field; he was by reputation an irrepressible source of ideas, many of them wild, but some highly original. His approach to research, in common with many restless minds, seems to have been one of smash-and-grab: scientists of this cast are easily beguiled by a clever idea. They quickly devise an experiment to test or partly test their conjecture, and if it works they move on in search of the next revelation, while others are left to sweep up the detritus. This manner of proceeding has its place in the ecology and its practitioners can be an inspiration, as much as an irritation, to their colleagues.

Stanley Pons was Fleischmann's *protégé.* He is an American from the deep South, who had abandoned his doctoral research to enter the family business, but had returned to science ten years later and completed the work for his Ph.D. thesis in Southampton. He was thus a late starter, but had quickly secured a faculty position at the University of Utah. He and Fleischmann clearly had a close rapport and Fleischmann would spend a part of each year in Pons's laboratory in Salt Lake City. Pons inherited his style of research from his mentor. His students found him stimulating and admired him, but at least one was uneasy about his professor's excessive productivity, which appeared to him to reflect a driving ambition to succeed more than a striving after scientific truth.

Steven Jones had spent his life in Utah. He was a professor in the Physics Department of Brigham Young University at Provo, some fifty miles from Salt Lake City. Like almost all the students and faculty at the university he is a Mormon, and apparently a very different type of scientist from Pons and Fleischmann. Respected by his associates as a pillar of social and scientific rectitude, he was cautious and conservative, and not given to speculation or hyperbole. He had a long-standing interest in nuclear fusion, which predated that of Pons at the rival university.

The quest for controlled nuclear fusion was nothing new and indeed the US Government, especially, had been supporting research in this area at a very high level for many years. To force two hydrogen nuclei (or those of its heavier isotope deuterium) into contact, so as to effect their fusion into a helium nucleus,[1] requires a huge input of energy because of the charge repulsion between them. In the hydrogen bomb this is accomplished using a nuclear fission explosion. Controlled fusion has been achieved by an energy input in the form of laser beams, or by accelerating the deuterium nuclei and causing them to collide at huge speeds. The high speeds mean high temperature, and the Tokamak, as it is called, maintains a gaseous plasma at

The Tokamak: fusion research in the real world. The photographs shows the TFTR, or Tokamak Fusion Test Reactor, at the Princeton Plasma Physics Laboratory. The upper photograph is an exterior and the lower an interior view. Princeton Plasma Physics Laboratory.

100 million degrees centigrade. So far, the energy output of these fusion processes has, at best, only barely exceeded the immense energy input. In Britain a nuclear fusion project had been started not long after the Second World War. In 1958 the Head of the British Atomic Energy Authority, Sir John Cockroft, announced that fusion of deuterium on an economic scale was within sight, but this quickly proved to be a mirage.

Fleischmann had been toying for some ten years with the notion that fusion might occur at an electrode at which deuterium was being liberated and had discussed the idea with his colleagues. Nuclear physics was not his field of course, but this did not inhibit his enthusiasm. His scheme was to liberate deuterium from heavy water (D_2O) at a palladium electrode. Palladium is one of the so-called noble metals, very similar to platinum, which is not dissolved by acids, and so the electrode is not corroded away. More especially, it had been known for many years that palladium will absorb large quantities of hydrogen, so that a high concentration builds up within the metal. Fleischmann's conjecture was that these circumstances would be favourable for fusion. This was in fact a less than original scheme, for as early as 1926 a well-known German physicist, Fritz Paneth, had reported fusion of hydrogen to helium, when absorbed in powdered palladium. (At that time there was a demand for helium as a safe alternative to hydrogen for lifting airships.) After further work it turned out that the helium, detected spectroscopically in minute amounts, came from the glass apparatus in which it was trapped and from which it was liberated only when hydrogen was present. The original claim of fusion was retracted the following year. Paneth's paper had excited much interest, however, and a Swedish physicist was inspired to try electrolysis; he also believed that he had formed helium at the palladium electrode. When later deuterium in the form of deuterium oxide (heavy water) became widely available, two Hungarian brothers, Adalbert and Ladislaus Farkas, working in the Department of Colloid Science in Cambridge in about 1935, thought they might have achieved fusion when they passed deuterium gas through a palladium tube. They showed their data to Lord Rutherford, the highest authority in the land on matters of atomic physics. Rutherford was sceptical, but agreed to look for the neutron release that must accompany fusion; none was found.

If Fleischmann knew of these early failures, he was not deterred, but it seems to have been 1984 before he and his disciple, Pons, went to work. They prudently desisted from applying for a grant to pursue what would have appeared to almost any reviewer as a crackbrained scheme, and they evidently invested a modest amount of their own money in purchasing the necessary materials. The apparatus was in fact so rudimentary that it later caused mirth at the atomic energy laboratory at Harwell, near Oxford, where the British interest in nuclear fusion was centred. It consisted in essence of a jar, filled with

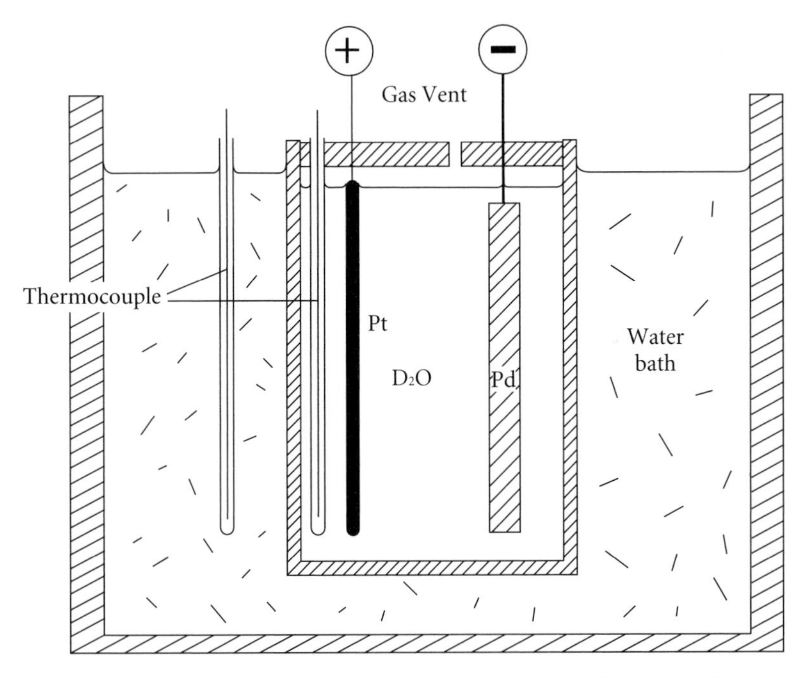

The cold fusion apparatus, depicted schematically. It was essentially a jar, containing heavy water and a dissolved electrolyte (lithium deuteroxide). The anode is platinum wire and the cathode a palladium rod, which sequesters the deuterium (heavy hydrogen) released by the electrolysis reaction. The thermocouples measure the temperature. The jar is immersed in a heat-insulated water-bath (a calorimeter), the change in temperature of which measures the heat output from the fusion cell.

a solution of an electrolyte[2] in heavy water, immersed in which were two metal electrodes—a platinum anode and a cathode (at which the heavy hydrogen would be released) of palladium. The electrodes were connected to a power supply and the assembly was placed in a heat-insulated water-bath, or calorimeter, so that by monitoring the temperature of the water one could determine the energy generated in the electrolysis cell. Knowing how much current went in, it would then be easy in principle to see whether more heat came out than could be accounted for by electrical heating alone.

The results were less than dramatic, but Pons and Fleischmann thought they were detecting a small excess heat output; and then one morning (it is not altogether clear how long after), when Pons's student went to inspect the cell, which had been running overnight in the basement of the chemistry building at the University of Utah, the apparatus had dematerialised. Fragments of glass were everywhere, one of the palladium electrodes appeared to have largely vaporised and a four-inch-deep hole had been gouged in the concrete floor. Perhaps explosive fusion had occurred? The episode was tendentiously referred to as a 'melt-down'. Explosions, though, are a familar hazard to those who habitually work with batteries: it is no secret that hydrogen and air are a potent mixture. And if the explosion had been the result of nuclear fusion the flux of neutrons that would have accompanied it would have been lethal to anyone in the vicinity. Worse, the neutrons passing through the water in the calorimeter bath would have given rise to high-energy gamma-rays—an even more alarming hazard. Pons and Fleischmann seem to have viewed the threat to life of a successful experiment with remarkable composure.

By 1988 Pons judged that the time had come to apply for financial support for the fusion project. His first thought was to ask the Office of Naval Research, which had already sponsored projects in his laboratory, but fearing that a positive outcome might become classified information, he changed his mind and turned instead to the Department of Energy, and in particular a body called the Office of

Stanley Pons (left) and Martin Fleischmann in the laboratory.

Advanced Energy Projects. This bureau disposed of an annual budget of $10 million, a sum that appears to have been in the gift of the director and sole administrator, a lapsed physicist by name of Ryszard Gajewski. By ill-chance, as it was to turn out, Gajewski's office was already supporting a nuclear fusion project: Steven Jones at Brigham Young University had developed an interest in fusion while working at the Idaho National Engineering Laboratory, where he had pursued a discovery made twenty-five years before by the particle physicist and Nobel Laureate, Luis Alvarez. This was muon-induced fusion, detected in a device called a bubble-chamber, which allows observation of individual reactions between subatomic particles. The muon is a heavy electron, two-hundred times the mass of an ordinary electron, with a transient existence. In liquid hydrogen (at –250 degrees) this particle can be captured during its brief lifetime of 2 microseconds by a hydrogen molecule (H_2, or when the bubble chamber contains a mixture of common and heavy hydrogen, by D_2, or the hybrid molecule, HD). The mass of the muon enables it to shield the charge repulsion between the two nuclei much more effectively than an electron, and an occasional spontaneous fusion event results. The muon is ejected when this happens and can then provoke another fusion. Alvarez later confessed that he had briefly wondered whether he might not have 'solved all the fuel problems of mankind for all time', but the process was too inefficient and the cost of generating a significant muon flux too great to encourage lasting optimism. Muon-induced fusion found its place among the curiosa of particle physics.

All the same, Gajewski backed Jones to the tune of $1.3 million for three years, and Jones appeared to make some progress in increasing the fusion yield (the number of fusion events per muon), although some work by a Swiss laboratory had concluded that the ejected muon would in general be caught by nuclei of the fusion product, helium, rather than bringing about any further fusions. At the same time, however, Jones—aware that, short of some dramatic development, he could not expect Gajewski's support for muon-induced fusion to continue indefinitely—began to think about another scheme. This he called piezonuclear (meaning pressure-induced) fusion. The idea, based on some work published in Russia, was to develop a high concentration of deuterium or a hydrogen–deuterium mixture in metals. For example, a deuterium-saturated metal foil might be subjected to high pressure in a diamond-tipped press, or a wire of the metal could be explosively vaporised by passage of a very high electric current. Among the metals that Jones and his group thought about was palladium, for they were also aware of its large capacity for absorbing hydrogen. Jones later had the jotting in his notebook, referring to palladium, notarised, lest he be accused of stealing the idea from Pons and Fleischmann (as indeed he duly was).

Fusion was to be detected, as in the muon experiments, by the release of neutrons, though only the most rudimentary apparatus was available in the department at that time. Jones accordingly prevailed on a neutron expert, J. Bart Czirr, to build a highly sensitive neutron detector to eliminate all doubts.

Not all of Jones's associates seem to have felt fully confident about his research strategy. Gary Taubes, in the most comprehensive and penetrating account of cold fusion (*Bad Science*), quotes the view of a junior research fellow who was entrusted with analysis of much of the data: Jones, he thought, was incapable of scepticism. His religious convictions precluded it, for he believed that God was leading him to the truth. Jones was also egged on by a confidant, an itinerant theoretician, Jan Rafelski. Gajewski had given Rafelski a grant of $3 million for seven years, even though an anonymous committee of leading physicists (the JASON committee), when consulted about the value of Rafelski's work, had turned their thumbs down.

When Gajewski received the application for support from Pons and Fleischmann he quite logically sent it to Jones and Rafelski for expert opinions. Jones must have been shaken to find a rival group so close at hand. He recommended rejection of the application, on grounds of various weaknesses, as he saw it, including an inadequate knowledge of the literature. In particular, no mention had been made of Jones's published work. Gajewski now tried to bring the two groups together. Why should they not make common cause and collaborate rather than compete with one another? Neither side was overjoyed by the suggestion. Fleischmann and Pons felt they were ahead of Jones and needed no help from his neutron detectors, while Jones was probably aggrieved that these interlopers were putting him under pressure. The grant application was eventually resubmitted and this time Jones made no objection. But soon after he had learned what was going on in Salt Lake City, Jones instructed a student to set up an electrolysis cell and look for neutron emission, without which, he rightly decided, the evidence for fusion would never be fully persuasive.

A new and unhealthy sense of urgency now took possession of all the participants, for, if cold fusion led to fame and patents, the first to publish would take all. Pons hastened to put his grant proposal before the university patents lawyer and redoubled his efforts to prove excess heat production in his electrolysis cells. A new graduate student, Marvin Hawkins, was taken on and bore much of the burden of tending the cells and analysing the data. A curious discrepancy as to the date of the 'melt-down' event entered at this point, for Hawkins described such a manifestation after he started work in 1988, whereas Pons and Fleischmann asserted that it had happened three years earlier. Perhaps there were two explosions? Yet the excess heat developed in the cells remained agonisingly small, if indeed there was any at all. Measurements that hover around the minimum level of detectability can be redeemed only by multiplying the number of experiments until the statistical error limits on the average magnitude of the effect become small enough. One might then find a 90 per cent probablity, say, that the effect is real, but to be driven to statistical wrangling of this kind would be regarded by many researchers as an admission of defeat. Pons and Fleischmann retained enough caution to wait for better data before committing anything to print.

Gajewski, meanwhile, was urging Jones to publish. Jones was due to give a paper on his muon-induced fusion at a meeting of the American Physical Society. Such presentations are accompanied by a brief published abstract, but this carries little weight compared to a peer-reviewed paper in a learned journal. Pressed by Gajewski, Pons and Fleischmann visited Jones in Provo. The Austrian satirist and critic Karl Kraus wrote that wars start when politicians lie to journalists and then believe what they read in the newspapers. Still using his primitive apparatus, Jones fancied he had detected a neutron flux from his electrolysis cell; it was at the very verge of detectability, but perhaps it was real. At all events, he would probably not have dwelt on the frailty of his data in the conversation with his rivals. The written accounts of the cold fusion story are at one in the surmise that Pons and Fleischmann were galvanised by what Jones told them:

if he had seen neutrons there was no longer any reason to doubt the validity of their shaky calorimetry.

Jones was persuaded to cancel a colloquium, which he was to have given in his department, but not his later presentation at the American Physical Society, the abstract of which was already submitted. A plan was tentatively agreed for both groups to publish their results at the same time, and if possible before Jones's abstract appeared. The President of the University of Utah, Chase Peterson, now intervened. He and his opposite number at Brigham Young University met in Provo, in conclave with Jones, Fleischmann and Pons. Peterson and his Vice-President for Research, Jim Brophy, had persuaded themselves that the discovery of cold fusion in Salt Lake City heralded a new era of prosperity for mankind, the United States and above all the State of Utah and its State University. Their counterparts at Brigham Young were startled, for, so far as they were concerned, the neutron fluxes did not suggest any great commercial possibilities, in fact rather the reverse. The University of Utah's representatives already had something of a reputation for excessive optimism. Two extravagant recent claims—one for an X-ray laser, which might have eclipsed the entire Star Wars programme, the other a workable artificial heart—attracted much ridicule when they came to nothing, and such ephemeral triumphs became known in scientific circles as the Utah Effect. An agreement was reached that two papers would be written and submitted concurrently to *Nature*: representatives of the two laboratories would meet on 24 March 1989 at the Federal Express office in Salt Lake City and so ensure that there was no skulduggery.

But Peterson forthwith broke the spirit of the agreement by calling a press conference in Salt Lake City to announce a historic breakthrough in energy generation, and, five days after their meeting at Provo, Fleischmann, Pons and the student, Marvin Hawkins, submitted a paper to the *Journal of Electroanalytical Chemistry*—some way downmarket from *Nature*. Fleischmann seems to have evinced some distress at being caught up in this stampede, for he must have known how insecure the claims of fusion still were. The pace, however, was being forced by the university's patents lawyers, since in American law patent rights are based on the earliest public disclosure of a prototype device. The applicant is then allowed one year in which to file his patent. Because Jones's impending presentation to the American Physical Society, with its published abstract, might serve as a disclosure, there was no time to lose. According to Taubes's account, Peterson himself was uncertain enough to seek reassurance from the highest possible authority. Hans Bethe, at Cornell University, was related to Peterson by marriage. He was then eighty-two years old, but his awesome intellect was undimmed. Bethe is one of the founders of modern theoretical physics, and played a critical part in the development of the atomic bomb. His reply to Peterson's inquiry was that cold fusion was highly unlikely. But, Peterson persisted, he had to consider the position of his university, in view of a similar claim, about to be published, from Brigham Young. Bethe rejoined that Peterson should let BYU publish and make fools of themselves. It was not what Peterson wanted to hear and he disregarded Bethe's advice.

The editor of the *Journal of Electroanalytical Chemistry* did not drag his feet. Editors of learned journals are often almost as eager as their authors for a publishing coup and the editor in question was no exception. He did not go so far as to circumvent the peer-review process, but the opinions of referees were quickly solicited. Their criticisms were passed on to Fleischmann and Pons and in some wise answered, and in a matter of days the manuscript was accepted and sent to press. The editor advanced the paper to the head of the queue and it appeared in print in four weeks, instead of in four months, as would have been usual. Peterson's press conference, and when it appeared, the paper, caused outrage in Provo. Brophy had even told the reporters that he was not aware of any work in the same area elsewhere than in Salt Lake City.

The press, led by the *Wall Street Journal*, did not delay, and the University of Utah was besieged by reporters and by excited well-wishers. The attraction of the story was obvious: the Tokamak and laser fusion projects were 'big science'. The annual cost to the taxpayer was $500 million, with little outward sign of progress, yet here were two insignificant chemists, who had invested a few thousand dollars of their own and achieved fusion in what looked like a jam jar. This was a quintessentially American romantic success story. The Board of Regents of the University of Utah was convened and addressed by the President, and it was at once resolved that the State Governor be asked for funding by his Legislature to launch a new institute for research into cold fusion. On the telephone, Governor Bangerter promised $5 million as a beginning.

The first the scientific community heard of cold fusion were the press and television reports. Photocopies of proofs of the article in *Journal of Electroanalytical Chemistry*—improperly released by the editorial office or purloined by a referee—then began to circulate. Jones now regarded the publication agreement as void and sent his manuscript to *Nature* directly by fax, where it was in due time accepted. Attempts to verify the startling result proclaimed by Pons and Fleischmann and announced in the press began almost at once. Pons received telephone calls from his *confrères* at other universities, asking for details of the cell, the electrodes and the electrolyte. Was the experiment dangerous? Was there a life-threatening neutron flux? It seemed not, but Pons had no explanation. Attempts to demonstrate fusion in the leading electrochemistry laboratories—at the Massachusetts Institute of Technology, the University of Texas, the California Institute of Technology and elsewhere—all failed. But all too soon, indeed before the *Journal of Electroanalytical Chemistry* paper appeared in print, reports of confirmation began to come in. The first two were from the Texas Agricultural and Mechanical University and the Georgia Technological Research Institute, and soon word came of successes in Japan, in Hungary and in Russia. The extent of the interest in the cold fusion phenomenon became apparent when Johnson Matthey, the world source of rare metals, ran out of palladium rods and the price of palladium rose vertiginously. One explanation for the failures to reproduce fusion effects was that not all palladium samples worked, and there was talk of 'fusion-grade' palladium.

The furore in the press rapidly communicated itself to the United States Administration, and the Department of Energy instructed several national laboratories to divert part of their effort into cold fusion. It also ordered a panel of experts to be set up to study the matter. The Committee on Science, Space and Technology of the House of Representatives initiated hearings, at which Peterson and representatives of the State of Utah gave recklessly optimistic accounts of what had been achieved and the brilliant future that it heralded. Nothing was said of the gigantic discrepancy between the fusion energies claimed by Pons and Fleischmann and that recorded by Jones and his group.

When they read Pons and Fleischmann's paper, the experts were appalled. Two pages of errata to the eight-page paper followed in the next issue of the journal two weeks later (including, grotesquely, the name of the third author, Marvin Hawkins, initially omitted), but a succession of gross errors and lacunae remained. Worst of all, the reported heat output exceeded what the minute neutron flux indicated by no less than eight orders of magnitude; it was too big, that is to say, by a matter of about one-hundred million. This inconvenient circumstance was lamely explained in terms of 'a hitherto unknown nuclear process'. The authors of the paper were accused of egregious carelessness, of elementary errors and of concealing essential details. A second paper, despatched to *Nature*, was rejected on grounds of both technical inadequacy and adding nothing to the first.

Yet the cold fusion story could certainly not be ignored. Fleischmann and Jones gave their very different accounts of the heat and neutron measurements at a conference of physicists in Italy, and, at the spring meeting of the American Chemical Society, a special symposium on the subject was hastily arranged. Pons spoke and was asked some searching questions from the floor. What about the obvious control experiment of filling the cell with light instead of heavy water? Surely this must have provided a baseline for all that was going on in the cell, besides fusion? 'A baseline reaction run with light (ordinary) water is not necessarily a good baseline reaction', Pons replied. Pressed to enlarge on this Delphic utterance, he was forced to concede that light water did not give a zero result; one should put one's faith in the evidence of calorimetry. Yet despite this and the huge discrepancy between the supposed heat output and the neutron release, the majority of the audience seem for inscrutable reasons to have accepted the reality of the phenomenon. A spate of lightheaded theorising now ensued, as Pons and several other physical chemists sought to explain the inexplicable. Somehow much of this got into print, editors perhaps overruling the opinions of the referees. At the Massachusetts Institute of Technology, a theoretician published four papers of supposed explanation and the Institute actually tried for a patent in competition with the University of Utah.

The physicists, on the whole, were not as easily persuaded as the chemists, and only a week after the American Chemical Society meeting the American Physical Society convened in Baltimore. Fleischmann had been invited to address the physicists, but prudently declined. The speaker who introduced the fusion sessions began with a curious point: of the now quite numerous attempts to repeat the cold fusion experiment, the laboratories of northwest Europe and the eastern and western seaboards of the United States, with one exception, had reported failure. These were the locations of the largest, richest and best-equipped laboratories, in which interdisciplinary investigations could most effectively be undertaken. The outcome of the most comprehensive study of all, carried out at the California Institute of Technology (the famous Caltech), was unfolded in the same session by Nathan Lewis. Experts in particle detection and electrochemistry had contributed and all possible variables—the form of the electrodes, the geometry of the cell, the time of charging, the current density and the electrolyte composition—were taken into account. Heat, neutrons, gamma-rays and helium were measured. The upshot was that no fusion could be detected, and certainly not at the level claimed by Fleischmann and Pons.

The only positive report came from R.A. Huggins at Stanford University, who had measured only heat output, but his work immediately came under fire from one of his own colleagues, who identified design faults in the calorimeter, which applied equally to the apparatus used at Utah. A leading theoretician, Stephen Koonin, explained that the supposed fusion yields would have violated the principles of such reactions. He accused Pons and Fleischmann of 'incompetence and perhaps delusion', and his peroration was greeted by applause. Steven Jones also spoke and was treated more gently, in proportion to the greater modesty of his claims.

A week later there was another meeting, that of the Electrochemistry Society, the tenor of which was very different. Pons and Fleischmann felt safe here and gave their contribution. The organisers decided that the emphasis would be on positive results, and it was with difficulty that Nathan Lewis persuaded them to let him speak. Participants from India, from Italy and elsewhere all reported fusion, as did many from the United States.

The criticisms, which became more numerous in the weeks that followed and were now given prominence in the press, discomfited Chase Peterson and his cronies in Utah. The attacks, Peterson said, came from large laboratories with a stake in the hugely expensive conventional fusion programme, or from the 'Eastern Establishment', which would never acknowledge a challenge to its primacy. Pons, meanwhile, was trying to hold at bay demands to allow his used palladium electrodes to be examined for helium content—another and sensitive criterion for fusion. He had made, he said, his own arrangements for an analysis—later reported, then withdrawn.

There followed yet another meeting, the strangest of all. It was sponsored by the Los Alamos National Laboratory (the centre for nuclear weapons research) and held in Santa Fe in New Mexico, and nuclear physicists, chemists, venture capitalists and representatives of the power companies and of Government were invited. The organisers, believers all, resolved to exclude all those with only negative results to report. In the vanguard was a veteran electrochemist, John

O'M. Bockris, an Irishman working at the Texas Agricultural and Mechanical University (the Texas A & M, as it is generally known), who many years before had taught Fleischmann in London. Bockris and others made extravagant claims of fusion efficiencies, far greater even than those seen by Fleischmann and Pons; no two laboratories agreed as to experimental conditions and the relation of heat to output of fusion products, and ludicrous theories to explain some or all the effects abounded. It was generally agreed that the results were irreproducible, some cells giving abundant fusion, others none, some working one day and not another, and that no laboratory could be expected to see the same results as any other. This was a very nadir of science.

In the respectable scientific literature more disquieting features began to emerge. Jones's paper in *Nature* reported a neutron flux of two neutrons per hour, a vanishingly small effect, but in their published paper Fleischmann, Pons and Hawkins had recorded a much higher amount of radiation. Because they could not count neutrons with any assurance, they had chosen instead to measure gamma-ray emission, in the first place that of the secondary gamma-rays resulting from the passage of neutrons through the water in the bath around the cell. When a neutron strikes a hydrogen atom in a water molecule, a nuclear reaction occurs, which is accompanied by the emission of a gamma-ray. The energy of the emitted gamma-ray is a signature of the reaction, and reveals itself as a peak at 2.22 MeV (million electron volts) in the emission spectrum. Now the paper included a picture of the peak, but it was shown without the rest of the spectrum on either side. This is a little like demonstrating that Venus is the brightest body in the night sky by showing a photograph of Venus by itself. In fact, there was just enough of the energy spectrum in the figure to disclose that another part of the signature, a satellite peak, resulting from collision of the gamma-ray with an atomic electron, was missing. Furthermore, the main peak was sharper than the power of the rather rudimentary gamma-ray detector to discriminate small energy differences could have allowed; and finally, the intensity of the peak was fifty times too small for the claimed level of neutron emission. So an expert reading the paper could see at once that the data were fishy, and in a critical letter to *Nature* by R.D. Petrasso and a group of colleagues at MIT the anomalies were put down to a defect in the detector. But this was not all, because in the errata to their original paper Fleischmann and Pons had shifted the gamma-ray peak to 2.5 MeV without explanation. They seem to have thought (mistakenly) that this was the correct wavelength for gamma-rays emitted along with the neutrons, and in a rejoinder to Petrasso's strictures they showed a complete spectrum, as demanded. This did not help, for as Petrasso's group at once perceived, the calibration standards, needed to establish the true energy, had been wrongly measured. Not only, then, were there no grounds for expecting an energy of 2.5 MeV, but the supposed peak shown in the spectrum was actually at 2.8 MeV. The whole spectrum seems to have been a mysterious artefact of the apparatus.

Fleischmann had meanwhile asked a friend in the Atomic Energy Research Establishment at Harwell, where there were excellent neutron-counting facilities, to look for neutrons directly. Cells were set up according to Fleischmann's prescription, but no neutrons were detected. In due course Fleischmann came to see for himself and was invited to give a lecture. Before a distinguished audience he told his tale, which was met by a good deal of criticism, not least of the spurious gamma-ray spectrum. His reception could have done little for his confidence and his statements to the press carried an unaccustomed tone of uncertainty. Soon after this, Harwell announced that research on cold fusion would not proceed.

Back in the United States, the Department of Energy panel went into action. They visited the University of Utah and the other laboratories playing a part, and found that where fusion had been recorded there was irreproducibility, wild variation of results, and confused and uninterpretable data. A report was promulgated, recommending that there should be no investment on a national scale in cold fusion research. There was some dissent, and two ageing Nobel laureates in physics, Julian Schwinger and Willis Lamb, sought to rationalise the phenomena by theoretical prestidigitation.

Bockris by now was the most vocal proponent of cold fusion and was making increasingly fantastic claims of heat output, and an independent colleague at Texas A & M, John Appleby, was confirming his data. The visit to the department of the Department of Energy panel proved a disaster. A subcommitte visited Appleby and soon discovered that he had little clue about what his two assistants, who ran the project, were about. Inspection of the raw data—the unprocessed recordings from the laboratory instruments—revealed no heat changes that could be distinguished from instrumental noise.[3] Appleby retired in confusion. The panel members could scarcely believe what they had seen.

Worse was to come when a second subcommittee visited Bockris's laboratory. One of the members, William Happer, Professor of Physics at Princeton University, described what he and his colleagues found: loose wires hung from the calorimeter that they inspected. Nothing was shielded and connections were made with alligator clips (little better than bulldog clips). Touching one of these caused the recorder pen, which measured the temperature, to jump several degrees. While injecting water into the calorimeter, the student in charge of the experiments touched the wire and the recorded temperature leapt. Such effects seemed to be taken as evidence for excess heat production. Happer then asked the student whether he ever observed negative heat changes. Oh, yes, the student readily conceded, there were negative effects aplenty and he brought out for the visitors' inspection rolls of recorder chart paper showing negative deflections of the pen. About these Professor Bockris did not wish to know, for it was only positive results that concerned him.

To convince the world that fusion was going on in the cells, Bockris next chose to pursue another criterion: an inevitable side-reaction of the deuterium fusion process would be the formation of the heaviest hydrogen isotope, tritium, in company with a proton. (Since two deuterium nuclei between them comprise two protons and two neutrons, and tritium one proton and two neutrons, a proton is left over.) Minute quantities of tritium occur in ordinary water and more especially in purified heavy water, and with sensitive counters these are not difficult to detect. Bockris's colleague, Kevin Wolf, who ran the cyclotron laboratory at Texas A & M, had found that prodigious amounts of tritium had appeared in some of Bockris's cells—far more than could be explained by any levels of fusion reported on even the most optimistic interpretation—but there was no sign of neutrons. Bockris had engaged in a show of chest-thumping, and instructed the Department of Energy panel that they had made fools of themselves. The tritium specialist on the panel, Jacob Bigeleisen, now confronted Bockris and Wolf with the question of where, if tritium had indeed been formed, the accompanying neutron had gone. Bockris, and more to the point, Wolf, had no answer.

From that moment the suspicion began to grow that Bockris's electrolysis cells had been spiked with tritium. Bockris hotly denied that this could be possible and submitted his findings to *Nature*, where they were promptly rejected as technically inadequate. The editor of the now infamous *Journal of Electroanalytical Chemistry* thought otherwise and accepted Bockris's paper, by-passing the process of expert review.

Wolf, who had grown increasingly uneasy at the paradoxes besetting the work, was now taking the possiblility of sabotage or fraud seriously, and quickly hit on a crucial test: to spike the cells with tritium, a miscreant would have to add tritiated water. Pure tritium oxide, T_2O, is not to be had and is available only in dilute form in ordinary (light) water, so if any tritium came from the outside an admixture of light water would be found in the heavy water in the cell; and so it proved. Wolf had had enough. 'The proper conclusion', he reported, 'is that things in the Bockris laboratory were so uncontrolled and so sloppy that those studies don't mean anything.'

That, one would have thought, would silence the egregious Bockris, but he counterattacked in one of the many letters with which he sought to browbeat the panel. 'We think', he declared, 'that, in attempts to verify a newly claimed phenomenon, negative results have much less value than positive ones. Negative results can be obtained without skill and experience.' It was the negation of all the principles of scientific probity. A colleague, who had worked independently to verify cold fusion, denounced Bockris's methods to the university administration and demanded an inquiry, but the university did not want to know.

Fleischmann and Pons now sustained what should have been a mortal blow. The panel had asked that the used electrodes be made available for independent tests for helium. Pons finally acceded to the demand and the palladium rods were sent, together with controls that had not been subjected to electrolysis, to six specialist laboratories on a double-blind (see p. 34) basis. No excess helium was detected in any of the used rods by any of the participating laboratories. A year later, in 1991, a paper from another laboratory was nevertheless submitted to the ever-eager *Journal of Electroanalytical Chemistry*, reporting helium emission in the electrode gases from an electrolysis cell, and was again accepted and published without review.

Equally troubling was the result of an offer by a member of the Physics Department at the University of Utah to help with the detection of fusion products. Michael Salamon suggested that he should be allowed to test the heat-producing cells. Pons agreed, and together with his team of nine research fellows and students Salamon set about the task of detecting neutrons, protons, gamma-rays and electrons, which should have come from another side-reaction. Nothing was found. Pons made excuses: the wrong cells had been examined, insufficient time had been allowed for the cells to become active, the detectors had been switched off just at the instant when a burst of heat had occurred. It was to no avail, and Salamon submitted a paper to *Nature* reporting all that he and his team had not found. No sooner had the paper appeared than Salomon and his co-authors, down to an undergraduate who had helped with the experiments, each received a letter from Pons's lawyer, who was being retained by the university, threatening the entire team with legal action if they did not immediately retract their statements in *Nature*. His own university refused to guarantee Salamon reimbursement of legal fees if the case should come to court, but Salamon nevertheless held firm and the matter was not pursued further.

At this point Pons resigned his chair at the University of Utah, and he and Fleischmann departed, to surface eventually in the South of France, where, with support from Japanese interests, they set up a laboratory. The University of Utah had meanwhile established its National Cold Fusion Institute with Governor Bangerter's $5 million, and there proceeded from it and from numerous laboratories around the world more positive reports, all of them vague or confused and mutually contradictory. Annual conferences on cold fusion were instituted, at which the fusiophiles could talk to each other, most reputable laboratories having by then lost interest. At the Third Annual Cold Fusion Conference in 1991 there were ugly scenes when a dissenter summarised the statistical outcome of cold fusion reports around the world. He was interrupted and abused by an Italian physicist and booed by the audience, and the microphone was snatched from him by the Japanese chairman.

Attempts to gather evidence for fusion by analysis of trace substances in palladium electrodes continued. Optimism rose when workers at the US Naval Research Laboratory announced that they had found what appeared to be a fusion by-product, a lithium isotope from the electrolyte. Edward Teller, a venerable figure in physics and in the politics of nuclear weapons, then in his eighties, suggested a new fusion mechanism, catalysed by an unrecognised subatomic particle, which he called the meshugganon (from the Yiddish expression for lunatic). But the lithium proved to be yet another mirage, the report was withdrawn and Teller evidently troubled himself no further with the matter.

A curious consequence of the collapse of the evidence for fusion from electrolysis cells was an upsurge in efforts to bring about fusion by equally improbable means. Fusion sightings were reported in high-pressure deuterium gas and in chemical reactions of deuterium compounds. Remarkably, the name of Boris Derjaguin reappeared.

He had not, it seems, learned caution from the debacle of polywater (Chapter 4), when in 1986 he announced that local fusion accompanied fracture of crystals of deuterium compounds. In 1992 he renewed this claim in another paper, but neither this, nor any other reported results from Russia and elsewhere, could be reproduced in American laboratories.

The National Cold Fusion Institute was wound up in 1992 after the Governor's $5 million had been consumed and silence reigned at the University of Utah, which, in the words of Nicholas Wade, science editor of the *New York Times*, 'could now claim credit for the artificial-heart horror show and the cold fusion circus, two milestones at least in the history of entertainment if not of science'. Cold fusion research had already migrated to the very fringes of science, especially to industrially funded laboratories in Japan. A new mania took hold among some of the wilder-eyed enthusiasts: the light-water cells had been reported in some hands to generate more heat, therefore more fusion, than the heavy-water cells, for which they were designed to provide a zero-fusion baseline. Such activities continue to this day, and reports are published in special journals.

Bockris had meanwhile enlarged his activities to encompass transmutation of heavy elements. He had transformed mercury into gold—a feat of which the alchemists had dreamed—but doubt was cast on this achievement when the research student carrying out the experiments was arrested and imprisoned for fraud. Bockris was not dismayed and went on to describe other successful transmutations. This proved too much for his colleagues on the Texas A & M faculty, who finally manged to have him removed.

The Pons and Fleischmann partnership broke up with the closure of their laboratory in France, where Pons apparently still lives in seclusion. They lost a lawsuit against an Italian newspaper, which had accused them of deliberate fraud—not an allegation ever made by their scientific adversaries. Cold fusion almost certainly inspired a play by Stephen Poliakoff, mounted by the National Theatre in London, called *Blinded by the Sun*; it centred on a scientific deception and the faked experiment was supposed to have achieved cheap energy generation.

Cold fusion is now a distant rumble and no longer disturbs the calm of nuclear physics or electrochemistry. The Seventh International Conference on the subject took place in April 1998 in Canada, and more than two hundred attended. The respectable scientific journals no longer consider manuscripts in this area. The catch-phrase is now 'chemically assisted nuclear reactions', CANR, and all manner of transmutations are featured. Some of its partisans are academics in respectable seats of learning, but a larger proportion seems to be members of small companies, which venture capitalists are evidently still backing to come up with an infinitely profitable revelation. Their psychology is that of the gambler on the national lottery, who knows he will not win, but thinks he just may.

What moral can one draw from this curious tale? In the first place, it is highly unlikely that experienced research workers like Fleischmann and Pons would have published, much less announced to the press, that they had already achieved cold fusion without the intense pressure of competition and the misguided demands of the President and lawyers of the University of Utah. Jones, similarly, would undoubtedly have wished to pursue his ideas at his own pace before risking his reputation by a premature announcement. It is entirely possible that a year or two of careful research would have persuaded all parties that they had not after all found a route to cold fusion. Their actions, then, were driven by political not scientific imperatives. Harold Macmillan, when prime minister of Britain, was asked by a journalist what were the main problems of his job. 'Events, dear boy', was his reply. The furore that followed the first announcement of cold fusion is harder to rationalise or extenuate. The show of unbridled ambition, venality and lack of principle displayed by a number of scientists around the world was shaming to the profession, and its consequences still reverberate.

Notes

1 The nucleus of the hydrogen atom is a single, positively charged, proton. In deuterium, or heavy hydrogen, the nucleus also contains a neutron (an uncharged particle with the same mass as the proton). There is another hydrogen isotope, tritium, which is radioactive, that is to say unstable and prone to disintegrate, and contains two neutrons in addition to the proton. When two nuclei of deuterium fuse, energy is released and a nucleus containing two protons and two neutrons is formed; this is the nucleus of the rare gas, abundant in the sun, helium.

2 An electrolyte is a substance which dissociates into positively and negatively charged constituents, or ions, when dissolved in water. A familar example (p. 75) is common salt, sodium chloride, written NaCl, which gives rise, when dissolved in water, to equal numbers of positively charged sodium ions, Na^+, and negatively charged chloride ions, Cl^-; the ions carry an electric current by migrating respectively to the cathode and to the anode. The electrolyte used by Fleischmann and Pons was lithium deuteroxide LiOD, an analogue of common alkali, sodium hydroxide, NaOH. The solution containes Li^+ and OD^- ions.

3 Any electronic instrument is subject to random fluctuations, for no device is ever perfect. Even the output of a battery for example, if measured with a sensitive voltmeter, will show a small up-and-down variation. This is known as noise. If one is measuring a quantity, such as the voltage output from a thermocouple, recording temperature, the noise will not even be noticeable if the voltage is large, and the recorder output will then appear smooth. If, however, one is measuring a voltage so low as to be comparable with the instrumental noise, this noise will be very much in evidence and will limit the precision with which the output can be measured or of the magnitude of the smallest output that the instrument is capable of detecting.

CHAPTER 7

What the Doctor Ordered

Lewis Thomas, the most lucid of writers on medicine and doctoring, called his discipline 'the youngest science'. Modern medicine assuredly has all the trappings of a science, and it undeniably makes abundant use of discoveries in the biological sciences, but it remains perilously poised between science and 'healing art', based as much on superstition as logic. The training of medical students still does not predispose to scepticism or an open mind in the face of text-book authority. The stance that doctors take on controversial issues commonly owes more to what they imbibed from their mentors than to a critical evaluation of the evidence. A debate, for example, has continued over two decades about whether high salt intake and hypertension are cause and effect. The argument in favour of the proposition appears to have been ruled by passion more than reason. The fog may now at last be clearing to reveal that there is no simple relation, and that millions of lives will not after all be saved by a low-salt diet.

The problems may be intrinsic to a profession forced to grapple eternally with the anfractuosities of human nature and the limited options for experimentation. One might ask, all the same, why centuries elapsed before doctors finally and reluctantly accepted what most laymen had known all along, that formulaic treatment of ailments—all the many designated, after Galen, as 'sanguine' in character—by bleeding killed rather than cured. As late as the mid-nineteenth century the lancet and the leech were still inseparable from clinical practice: in a single year around this time more than thirty million leeches were imported into France alone. Such blinkered adherence to doctrine did not end with the advent of 'scientific medicine', in the age of Osler and Treves. Consider for example the history of 'dropped organs', also known as ptosis. The belief grew in the late nineteenth century that all manner of symptoms arose from sagging innards—the stomach, the viscera and, most of all, the kidneys. Anne Dally, in her writings on the subject, has suggested that the misconception may have arisen from the experience of the dissecting rooms, where the offal inside the cadaver had been made rigid by the preservative, formaldehyde. If this is so, then observation in the butcher's shop might have served the doctors better. When X-rays came to their assistance, doctors were confirmed in their diagnosis of sunken or mobile organs in the standing patient.

The kidneys, at least in skinny people, could be felt, and the diagnosis of a dropped or mobile organ instantly confirmed. The kidney, being a loosely floating body, proclaimed the acuity of the doctor's diagnostic skills even more triumphantly when the patient was opened up. One school of thought distinguished between 'floating' and 'movable' kidneys, which were either insecurely attached to a 'mesonephron' (or 'side-kidney'), held to be a fold in the peritoneum, or were displaced to a position behind the peritoneum. Abdominal operations were in the heyday of ptosis highly hazardous, and the most radical intervention, thought to have begun in 1883, was to cut into the peritoneum and take out the kidney. The death rate, presumably mainly from peritonitis, was about 50 per cent. A German surgeon developed the slightly less heroic measure of sewing the kidney capsule to a muscle of the abdominal wall. This did not require perforation of the peritoneum and, as a less dangerous procedure, was eagerly adopted and given the name of nephropexy.

Many thousands of these useless operations were performed, the greater part on women, who were held to be more susceptible to the condition. A monograph on the subject of the movable kidney, published near the beginning of its era, had already determined that there were no recorded deaths from the condition, but no small number from the operation. An alternative treatment for ptosis, which enjoyed a brief vogue, was to apply tight bandages to the abdomen to hold the organs in place. But this was eventually judged ineffective and the surgeons were allowed to have their way. Symptoms viewed as indicators for surgery included abdominal or back pain, some heart and menstrual problems, persistent nausea, halitosis and constipation, and also neuroses. A French doctor identified ptosis as a cause of 'neurasthaenia'—a nervous debility or state of lassitude, beloved of Freudian psychiatrists—and a new syndrome, Glenard's disease, was born.

Nobody seems to have questioned the reality of ptosis for some fifty years, but the medical profession took a grip on itself in the late 1920s, when doubts were at last voiced. Over the next two decades ptosis quietly slipped out of the text-books, but in some centres floating kidneys were still being surgically corrected in the years following the Second World War. Ann Dally reports sightings of dropped organ diagnoses in 1992, however, so one cannot be sure that patients are not even now being subjected to this highly unpleasant ordeal. Tonsillectomy, to which untold numbers of children were subjected during the first half of the twentieth century, was fortunately in general without overt ill-effects, and so its efficacy was barely questioned until the mid-1950s, when the practice gradually fell out of favour. The same was true of uterine 'dilatation and curetage'.

Surgeons were always easily persuaded of the benefits of removing dispensable body parts. George Bernard Shaw satirised the practice in *The Doctor's Dilemma*, in which the surgeon, Cutler Walpole—commonly thought to be based on Sir William Arbuthnot Lane, of whom more shortly—finds an inessential organ, the nuciform sac; this he gives out to be the seat of every kind of disease and earns social and pecuniary success by relieving his grateful patients of it. In the years before Shaw wrote the play surgeons had for instance been taking out healthy ovaries from untold thousands of women as a cure-all for undiagnosed conditions. This operation at the time had a mortality rate of some 20 per cent.

Soon thereafter a new fad began to infiltrate medical thinking. That diseases were caused by accumulation of the body's own toxins was not a new notion,[1] and it had indeed been the origin of the spas which sprang up around Europe in the early nineteenth century. Drinking the foul-tasting sulphurous aperient waters, or sometimes taking them in enema, relieved constipation, real or fancied, and purged the body of its malign waste-matter. The most fashionable spa of them all, Baden Baden, was said to have been built on undelivered faeces, and in France Plombières gave its name to the 'Plombières douche', also known by the euphemism of *lavement*; in England it was 'intestinal lavage' and later 'colonic irrigation', while in the United States 'colon laundries' drew a dedicated stream of hypochondriacs. Expressions such as chronic obstruction or stasis, auto-intoxication, putrefaction and faecal sepsis, appeared with increasing frequency in the medical literature. Shaw absorbed the jargon: Cutler Walpole's nuciform sac (or rather that of his patients, for he himself was fortunate in belonging to the small fraction of the population not burdened with one) was stuffed with 'rank ptomaines'.

The theory was given respectability by Élie Metchnikoff, one of the founders of immunology. Metchnikoff was a Russian, who spent much of his working life at the Pasteur Institute in Paris. His fame rested largely on an inspired discovery—that of phagocytosis, an essential element in the body's defences against infections. Phagocytes are cells that engulf and digest invaders, such as bacteria, and destroy dead or moribund cells. Hereditary conditions that inactivate phagocytic cells are generally fatal. Metchnikoff made other contributions to immunology, but as he grew older he became increasingly preoccupied with the mechanisms of senescence and the possibilities of prolonging life. He expounded his ideas in many publications, but the most influential was a book, *La Vie Humaine*. An English translation by a prominent biologist and admirer, P. Chalmers Mitchell, appeared to wide acclaim under the title *The Nature of Man—Studies in Optimistic Philosophy* in 1903. Metchnikoff believed that diseases, among which he counted the manifestations of old age, were caused by toxins, generated in the large intestine by bacterial fermentation. 'The large intestine', he

Élie Metchnikoff. Science Photo Library.

wrote, 'must be treated as one of the organs possessed by man and yet harmful to his health and his life. The large intestine is the reservoir of the waste of the digestive processes, and this waste stagnates long enough to putrefy. The products of putrefaction are harmful. When faecal matter is allowed to remain in the intestine, as in cases of constipation, a common complaint, certain products are absorbed by the organism and produce poisoning, often of a serious nature.' He then goes into some of the conditions, from acne to stomach cancer, that these processes may occasion. 'In fine', he sums up, 'the presence of a large intestine in the human body is the cause of a series of misfortunes'. It was evidently a failure of evolution not to rid us of this succubus. But parrots, Metchnikoff notes, are exceptionally long-lived birds; and the reason? Why, because they have short guts, and to shorten the gut is to lengthen the life.

Much of Metchnikoff's book is given over to an analysis of ageing. There is warfare between our 'higher elements' and the 'simpler or primitive elements', and when these last win, we die. Metchnikoff identifies these injurious elements as his famous phagocytes. They may rally to our defence when microbes invade our bodies, but in time they come to feed off our tissues; and so 'atrophy of the kidneys' is caused by macrophages, and 'in the brains of old people a number of nervous cells are surrounded and devoured by macrophages'. He then offers us a truly gruesome vision: the hair, he informs us, consists of two concentric layers, throughout which its pigment is distributed. But there comes a moment when 'the cells of the central cylinder become active and proceed to devour all the pigment within their reach. Once they are filled with coloured particles, these cells, which are a variety of macrophage...become migratory, and, quitting the hair, either find their way under the skin or [let us hope] leave the body'. Bones become brittle in a similar manner, sclerosis and kidney atrophy begin their relentless progress and 'nervous centres' are eaten by the rampaging macrophages, in the manner also seen in general paralysis of the insane (otherwise known as tertiary syphilis). Old age, Metchnikoff pursues, warming to his theme, is pathological, for we do not wish to grow old, and 'an instinctive feeling tells us that there is something abnormal in old age. It cannot be regarded as a part of healthy physiological function'.

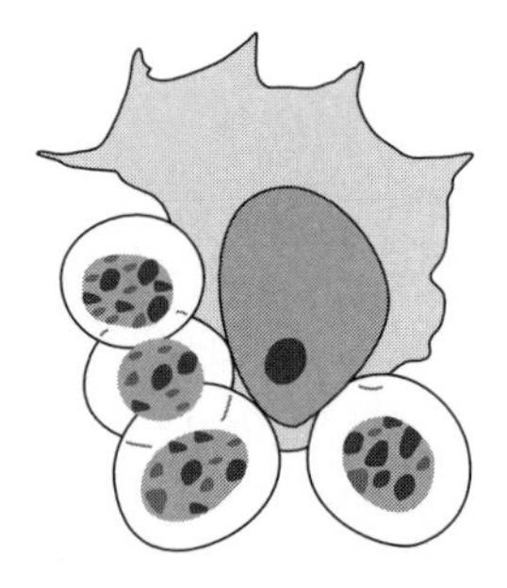

Illustration from Metchnikoff's book, *La Vie Humaine*, showing some of what he believed could be seen in the microscope: a drawing, captioned 'Cell from the brain of a woman 100 years old being devoured by macrophages'.

How then are we to avert this unnatural calamity? One option, it seems, is through small doses of antiserum (against what is not made clear), for in an experimental rabbit such treatment increases the number of red blood cells in the circulation and so 'reinforces the higher elements'. Better yet, though, is to combat the gut bacteria, which thrive on the putrefied matter and mucus. Milk, says Metchnikoff, putrefies only slowly and is protected by the bacteria that turn it sour. The acidity is due to lactic acid, and inhibits the multiplication of the gut bacteria, which grow only in alkaline conditions. To keep the toxicity at bay, therefore, we should consume much soured milk or *kephir* (something akin to yogurt). We must, moreover, avoid alcohol, which 'of course' diminishes the 'vitality of certain important cells', and eat only cooked or sterilised food, for otherwise we may ingest 'wild' microbes, especially from manured soil. If we follow these rules we will recapture the longevity with which the men of old were blessed—like so many biblical personages, Methuselah, Noah, Aaron, Moses, Joshua and Job among them. Then, as now, tales of hale and sexually active centenarians, living on yoghurt in the mountains of Asia Minor, circulated and were unquestioningly believed.

This extraordinary farago exerted a huge appeal and to the man who was to put Metchnikoff's teaching into practice in England, it came as an epiphany. Sir William Arbuthnot Lane was a highly respected and conscientious surgeon at St. Batholomew's Hospital in London. He appears not in fact to have been the model for Cutler Walpole, for Shaw later averred that he had not heard of Lane when he wrote the play, but the similarity is remarkable. Lane was a scholar, an evolutionist, predisposed to believe that the human body was littered with 'evolutionary relics', redundant and at best useless, at worst pernicious. He studied Metchnikoff's writings, and in due time had the transforming experience of making his hero's acquaintance. At Lane's house the two men talked and became friends. Before long Lane was preaching the doctrine of auto-intoxication and descanting on the virtues of frequent evacuation and on the evils that lurked in

Sir William Arbuthnot Lane. By courtesy of the National Portrait Gallery, London.

the colon (ideas that survive today in the cult of 'colonic irrigation', popular in certain high places). The colon was an evolutionary relic and we would all be better off without one. Constipation was the result in large part of man's counter-evolutionary upright posture, which caused the end of the intestinal syphon trap to fall and so obstruct its efflux. The loop would become distended and accumulate waste matter, mucus and blood with hideous consequences. These visitations were compounded by social inhibitions, which prevent us from seeking easement at the necessary frequent intervals. The condition of 'chronic intestinal stasis' was responsible, Lane asserted, for a swathe of ailments, ranging from bags under the eyes and bad breath to inflammation of the gall-bladder, thyroid disease, rheumatic fever and cancer.

Lane, having convinced himself, with Metchnikoff's encouragement, that constipation brought all these woes, looked around for a suitable treatment. His first suggestion was to drink a pint of cream daily as a lubricant; secondly corsets were proscribed but everyone should wear a tight belt to keep the abdomen as close to the chin as possible; finally Lane enjoined squatting on the floor, instead of sitting on a chair and also sleeping stomach-down. When none of these measures proved effective, Lane proceeded to treat constipation surgically. Excision of the colon is not a pleasant procedure, nor did Lane appear to trouble himself with follow-up reports. The number of these operations is known to have far exceeded a thousand. Inspection of his patients' exposed colons further persuaded Lane that they were subject to dangerous 'adhesions', and the features of colonic topography became known to his followers as 'Lane's First Kink', 'Lane's Second Kink' and so on.

Lane had begun, he wrote in 1908, by removing most of the colon only to relieve pain (from what is not clear). It was then borne in upon him that after such operations symptoms of 'auto-intoxication' would disappear. The change in his patients was immediate and 'almost startling'. This revelation started him on his messianic path. 'The recognition of the immense advantages which these miserable people obtained from the removal of the large bowel then induced me to operate also in cases where pain was not necessarily such a marked feature, but where life was becoming a burden through misery and distress induced by auto-intoxication and its results.' As time went on and the number of operations mounted, Lane came to see clearly that a prodigious variety of disorders had their roots in the colon. And 'Metchnikoff's work has naturally been of the greatest interest in the study of these changes'. Lane then went on to give details of the outcome of the operation on thirty-nine patients, not a few of whom died. He prefaced his account with the observation that he did not 'propose to attempt to prove anything by statistics, as personally I have very little faith in them'. An editorial in the same issue of the *British Medical Journal* noted that these statistics revealed a mortality of 18 per cent, but, it owned, some of the patients would have been 'reduced to extremities' before Lane applied the *coup de grâce*. The journal had also been commendably sceptical about Metchnikoff's prescriptions for immortality all along. An editorial discussing an account of work by Metchnikoff and Lane in 1904 sagely concluded: 'It remains to be seen, however, whether man's days in the land will be lengthened by the shortening of his gut'.

Soon thereafter Lane published a pamphlet, with the title 'Civilisation', in which he explained that modern man's upright posture and his deplorable custom of sitting on chairs, rather than squatting on the floor, caused the drop of the viscera, with its attendant evils. He went into the symptoms that 'toxic' people displayed. He spoke of changes in pigmentation and 'a nasty graveyard odour' that such unfortunates exuded. But, though he had a number of ardent disciples, he did not carry the weight of medical opinion with him. At a series of meetings of the Royal Society of Medicine in 1913 passions ran high on both sides. Lane, 'pale, tense and tremulous with emotion', according to an eyewitness, accused his critics of having closed minds. 'It is now too late for quibbling', he instructed them, 'and for the negation of facts which are perfectly obvious and capable of demonstration'.

Despite all criticisms Lane went on cutting out colons for another ten years or so. But in the meantime he continued to experiment with lubricants. After cream came olive oil and cod-liver oil, and then Lane found himself one night seated at a dinner next to an engineer. Could his neighbour think of any chemically inert lubricants that were not noisome or toxic? Why not paraffin oil then? Lane tried it on himself with gratifying results, and soon casks of the fluid began to arrive at the hospital like beer barrels at a pub. Flavoured emulsions appeared in the chemists' shops, and paraffin oil quickly became the sovereign treatment for a range of ills. Lane was in no doubt about its value: 'I think I may safely assert that no remedy has rendered so much good to the human race as paraffin'. He appears to have shared Metchnikoff's belief that, were it not for auto-intoxication, the effects of ageing could be averted. He visited the sinister French surgeon, eugenicist and fascist, Alexis Carrel, who, when working at the Rockefeller Institute in New York, had claimed (incorrectly, as it turned out) that animal cells would divide indefinitely in culture and were thus immortal. Was paraffin oil then the elixir of eternal youth? It was only after some years that patients turned up on the autopsy table at St. Bartholomew's Hospital with enlarged livers, from which the paraffin oil could be wrung as from a sponge. During the Second World War, when food shortages stimulated the search for substitute materials, the Ministry of Food recognised the danger and forbade the sale of salad dressings made with paraffin oil. Sir William Arbuthnot Lane died, laden with years and honours, in 1945.

Cutting out the colon had gone out of fashion years earlier, but surgeons still found it difficult to resist tinkering with the organ, given the least encouragement. A rational London surgeon, A.F. Hurst, fulminated against the practice in 1935. The position of the colon, he wrote, varied from one individual to another, and its tone depended on how full it was. It was often radiologists who made the diagnosis of 'ptosis of the colon' or a 'pelvic caecum', on which the surgeons then eagerly seized. He gave as an example of futile operations the case of a young woman, who at the age of seventeen had had her appendix removed to relieve abdominal pain. This was bad enough, but the pain had persisted, so two years later she underwent another operation to eliminate adhesions. It did not help either, and another four years on a different surgeon operated for adhesions on the colon. But, he wrote to Hurst, he had found none, though 'the bowel being "abnormally movable", he fixed it—thus doing exactly what he intended to undo when he advised operation. After this', Hurst continues, 'the symptoms were worse than ever, but they disappeared when the patient was persuaded to give up aperients, which she had throughout been taking in excess'. Auto-intoxication indeed! It is not entirely clear how long it took for the warnings of Hurst and a few other enlightened doctors to take effect.

Untold numbers of the sick and healthy have suffered over the years in the interests of surgical fads. For many years patients dying of tuberculosis were subjected to the gruesome pneumothorax operation, on the baseless theory that if the lung were collapsed, and thus 'rested', by cutting away a large portion of the rib-cage, it might spontaneously regain its function. And at the other end of the spectrum, it is not so long since a majority of children, at least from middle-class families in Western Europe and America, were made to undergo totally unnecessary tonsillectomies.[2]

The search for a means of prolonging life and especially maintaining youthful vigour, generally identified of course with sexual potency, has a long history. Injections of testicular extracts had been tried at various times, and this treatment underwent a resurgence at around the turn of the twentieth century. A distinguished French physiologist and pioneer of endocrinology (the study of glandular secretions, or hormones) Charles-Édouard Brown-Séquard, who had succeeded the great Claude Bernard to the Chair of Physiology at the Collège de France in Paris, treated himself with such extracts, not to mention dog semen, when he was already over seventy. He reported to a meeting of the Société de Biologie the remission of various symptoms of ageing, return of sexual prowess and even an increase in the measured trajectory of his urine. His account was met with general scepticism, but the evident efficacy (albeit transient) at about this time of implants of animal thyroid and adrenal glands in alleviating the effects of hormonal deficiencies, and Brown-Séquard's high reputation caused some physiologists to take notice. He and his associate, Arsène d'Arsonval (who, as we have seen (p. 10), was to delude himself again some years later over the supposed physiological effects of N-rays) promoted a system of 'organotherapy'—the treatment of a range of diseases with glandular preparations. An estimated 1200 doctors were persuaded of its efficacy and offered it to their patients.

An interest in the use of testicular extracts in particular, and in implants of animal testes into the elderly and infirm, began to take hold in Europe and in America. In the years immediately following the Great War, Eugen Steinach at the zoological research institute of the University of Vienna, the Vivarium, performed animal experiments, which purported to demonstrate the success of intra- or inter-species implants of testicle slices. At this time the phenomenon of rejection of grafts from one individual (or especially species) to another was imperfectly understood. The honour of being first to try the procedure on humans appears to belong to a physician in Chicago, but it was Steinach who made it famous. Many ageing libertines submitted to the operation, as well as to another Steinach speciality—ligation of the vas deferens, or sperm duct, in one testicle. The contingent retention of sperm was held to conserve virility and the surviving functional testicle sufficed to ensure fertility. Steinach's patients, at all events, appeared generally well-satisfied, though Sigmund Freud, who submitted to 'a Steinach' in the hope that it would help him combat his jaw cancer, could detect no betterment.

In Paris, meanwhile, another imposing figure was establishing himself in the testicular implant business. Serge Voronoff was a Russian-born surgeon, who had obtained a position at the Collège de France, and was trying out testis grafts on old and on castrated sheep, and reporting spectacular successes. The criteria were of course wholly subjective, and, as we now know, they must have been illusory, for the grafts would have been rejected in a matter of days. Voronoff, like Sir William Arbuthnot Lane, was influenced by the (spurious) observations of Alexis Carrel, a personal friend, on the survival of cells in culture, and believed that the fluid within the scrotum would similarly 'energise' the implant. Voronoff began work on human subjects soon after the First World War. For them he chose monkey testes

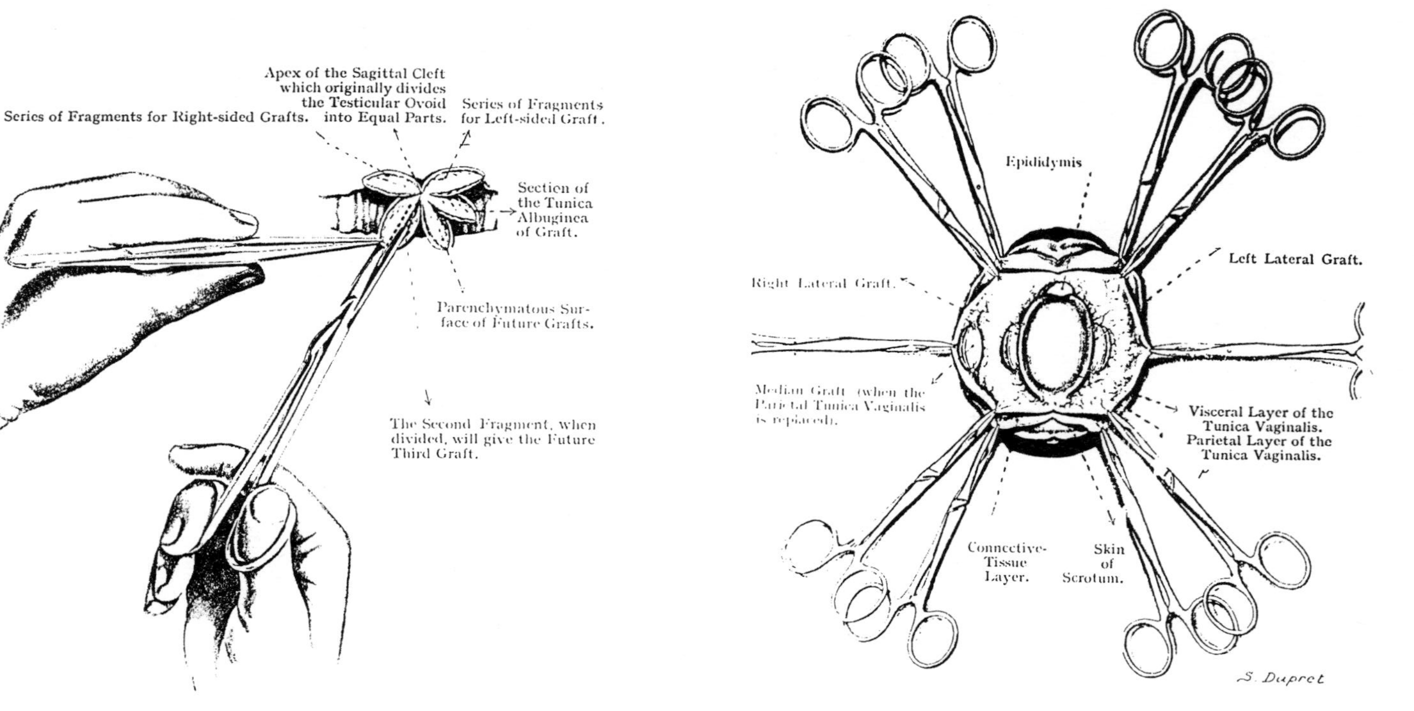

Voronoff's monkey testis implant operation. On the left we see the procedure for depriving the monkey of its possession, on the right the process of implanting it into the human recipient.

as implants, on grounds of biological compatibility. The operation was simple: the patient's testis was laid bare and its surface roughened to promote adhesion of the graft, and three slices of the unfortunate donor's testis were stitched in. His first two patients developed serious wound infections, and the grafts had to be excised, but in general there seem to have been no deaths from bacterial infections and no cases of what could have been lethal transfer of viruses from the monkey tissue. There were, on the other hand, many enthusiastic testimonials from rejuvenated beneficiaries. Within a few years, Voronoff had performed a thousand such operations.

The medical profession for the most part held its peace; muted disapproval was offset by uncritical acclaim in some quarters. In 1924, the respected *Scientific American* commented: 'Even death, save by accident, may become unknown, if the daring experiments of Dr Serge Voronoff, brilliant French surgeon, continue to produce results such as have startled the world'. Testis grafts were performed in England and elsewhere in Europe, but it was in the United States that fervour burned most brightly. In 1919 Dr Leo Stanley, the medical officer at San Quentin gaol, had removed the testes of an executed murderer and grafted them into an ageing and ailing felon, with purported success. This ghoulish exercise foreshadowed the trade carried on today in China, where executions of felons are numerous and are often, it is said, arranged to suit organ recipients, who fly in with their surgeons from the rich countries of the West. Stanley was encouraged to perform many more such operations, and especially testicular tissue injections on his captive population, possibly using material donated in exchange for privileges. In the mid-West, meanwhile, there rose up an ingenious quack, 'Doc' John Brinkley. He had no genuine medical qualifications, but offered and performed goat testis implants, in standard and *de luxe* versions, on his ranch in Kansas. He started a radio station to advertise his services, housed his patients in specially built hotels and sold them quack medicines. Brinkley became a national celebrity and received at one stage 3000 inquiries a day from anxious men (the sort of numbers now seeking the answer to their problems in Viagra). He ran for the governorship of the State and was only narrowly defeated. Brinkley's luck eventually ran out when his bogus medical degrees were exposed. As Dr Johnson said of the notorious eighteenth-century quack, John 'Chevalier' Taylor, his career was an illustration of how far impudence can carry ignorance.

Voronoff was convinced that testis implants could improve animal husbandry. He operated on sheep, bulls and race-horses. (In Britain the Jockey Club announced that stallions that had been treated in this manner would be struck from the stud-book.) In a large-scale study in Morocco, Voronoff had convinced himself that the wool-yield of rams was greatly increased after grafting, and also that this acquired character was transmitted to the offspring. This rank Lamarckianism— the discredited doctrine of inheritance of acquired characteristics, and the converse of Darwinian evolution—was in tune with beliefs current in Steinach's Vivarium, and especially the work of Paul Kammerer (Chapter 2), who in 1924 published a monograph on rejuvenation. Voronoff's claim incensed the veterinary scientists, whose grasp of biology and experimental design appears to have far exceeded that of the doctors, and a proper investigation of the matter revealed no discernible effects of testicular grafts on wool yield or any other feature of interest.

But Voronoff's eventual fall came from the discovery of hormones, and especially of the male sex hormone, testosterone, in 1929. It was shown to enhance secondary sexual characteristics when adminstered to animals, but it had no rejuvenating effects. Voronoff lived on until 1951 and died in obscurity. Steinach's ligation operation remained popular throughout the 1930s, and in Australia a lone surgeon continued to offer—with apparently no shortage of takers— the monkey testis implant operation until the Second World War, when the Japanese invasion of the Pacific put a stop to the supply of monkeys.

By this time another star began to burn brightly in the firmament of quackery. Dr Paul Niehans in Switzerland, once a follower of the Voronoff school of testis grafting, developed his 'tissue therapy', based on cells from sheep fetuses. Rejuvenation followed, and among Niehans's satisfied customers was Pope Pius XII (evidently anxious to defer the meeting with his superior) in 1953, and, it was said, the Duke and Duchess of Windsor, as well as countless lesser celebrities. Niehans was much the most acclaimed of many rejuvenators around the world, and no doubt, with a limitless supply of gullible customers, the trade continues still. For a vivid and authoritative account of Voronoff's life, and the background, scientific and social, of his work the reader should turn to David Hamilton's book, *The Monkey Gland Affair.*

The history of medicine is of course rich in fanatics and fantasists (quite aside from the outright charlatans), who have dunned, tormented and killed the gullible and vulnerable, but few started new fashions in treatment or achieved any but a narrow following. One of the most durable cults, which retains its hold in the rich countries of the West after two centuries, is homeopathy. This system was conceived by a German doctor, Christian Friedrich Hahnemann. His teachings were derided by the medical profession in his country, and were indeed absurd. Hahnemann held that all diseases were of four kinds—iatrogenic, caused, that is to say, by the inept ministrations of doctors, or derived from the 'psora', or itch, from syphilis, or from 'sycosis', a necrosis of the skin. Of these the itch was the commonest, and serious conditions resulted when it was internalised. The remedy for any disease was something that simulated the symptoms of the disease in weaker form. The homeopathists' saw was (and is) *similia similibus curantur*. This was not even then a new principle: the assertion that 'fire drives out fire' was known to Shakespeare (it occurs in *Julius Caesar*), and implies that burns were treated with heat. Alcoholics have always believed that a hangover is best combated with 'a hair of the dog that bit you'. A purgative, then, according to homeopathic doctrine, may be used to treat diarrhoea; atropine, which causes hot flushes and a high temperature, to treat fevers, and so on. Drugs, moreover, have 'signatures', revealed by the shape and colour of the plant or animal from which they are derived. So the root of the orchid resembles the testicle (as its name, from the Greek, *orkhis*, the testicle, implies) and it can therefore cure diseases of this organ); similarly the euphrasia, or eyebright, a flower with a black spot, reminiscent of the pupil, is a remedy for diseases of the eye.

But of course dangerous substances must be given only in innocuous amounts, and so preparations were diluted by successive factors of fifty, the solution being agitated, or 'triturated', at every stage to release the active essence, which became stronger with every successive dilution. It was no wonder that Hahnemann was hounded out of Leipzig by the enraged apothecaries, who wanted to sell drugs in the largest possible quantities. In the end, of course, the solution administered to the patient probably contained not so much as a single molecule of the substance: a nineteenth-century critic likened the treatment to pouring a cup of medicine into the Thames at Windsor and taking out a spoonful at Southend. The curative nostrum is in fact water.

In an age when doctors killed more often than they cured, homeopathy did no harm, nor does it still. It undoubtedly even does good, for what Hahnemann unwittingly discovered is the placebo effect—the improvement that results from the patient's faith in the remedy. (The etymology of the term—'I shall please'—is rooted probably in the Vulgate, where it occurs in a psalm; the passage translates from the Latin as 'I shall be pleasing unto the Lord'. Placebo came to be used for any trifle that gave pleasure or comfort, and entered medicine with the concept of controlled trials of new drugs.) It used to be said in the days before syphilis could be treated with Salvarsan or antibiotics that the patient had the choice of homeopathy or allopathy—of dying from the disease or from the cure. Like practitioners in all irrational cults, homeopaths have always resisted evaluation of their system by rigorous, controlled statistical trials. This is equally true of psychotherapists, for example. The beliefs of Sigmund Freud and his countless disciples, which coloured so much of the intellectual and artistic discourse in Europe and America (but is scarcely relevant here in the context of science), may now be undergoing a terminal eclipse, but homeopathy still thrives, most notably in France (Chapter 1), but also in other Western countries.

The antithesis of homeopathy—the administration by ignorant or insouciant doctors of dangerous substances in huge doses for trivial disorders—can be illustrated by many a grim episode in medical history. One of the most macabre was the brief vogue in the 1930s for drinking solutions of radium compounds as a cure-all and tonic. This may have stemmed from the discovery of the radioactive gas, radon, in the waters of European spas. Some doctors inevitably leapt to the conclusion that radioactive emanations were the health-giving principle. A dangerous quack with spurious medical qualifications, William Bailey, set up a company in the United States to market the Radioendocrinator, a bracelet containing high concentrations of radium and thorium, to be worn around the neck for the benefit of the thyroid, round the body to assist other organs, and, as one might expect, around the scrotum for failing sexual powers. Another of Bailey's companies marketed Radiothor, a solution of radium salts, advertised as a rejuvenating agent (notwithstanding the evidence, already available, of the hazards of radioactivity). Radiothor was prescribed by doctors and also sold over the counter in pharmacies, and probably some thousands of people drank it regularly. How many deaths it occasioned is not clear, but matters came to a head (see *The Great Radium Scandal* by Roger Macklis), when a well-known New York sportsman and socialite, Eben Byers, died in gruesome circumstances, his bones disintegrating, in 1932. He had consumed, over a period of four years, between 1000 and 1500 bottles of the elixir, many times the lethal dose. Bailey's business was quickly shut down. When Roger Macklis tested empty bottles of Radiothor sixty years later, he found that they still harboured frightening levels of radioactivity. Byers's bones and teeth blackened photographic plates, and the bones of other victims of Radiothor-induced cancer, exhumed in later years, were still too radioactive to be safely handled. Episodes of this kind reveal homeopathy in a highly favourable light.

The rise and continuing survival of homeopathy violates Langmuir's criteria for pathological science (Chapter 3). It is probably closer to a superstition than to an aberration of science, and is perhaps out of place here. An early parallel to the examples of transient delusions in the 'hard' sciences, is phrenology, or 'craniology'. This was invented by a Viennese doctor and physiologist, Franz Joseph Gall, at the beginning of the nineteenth century. It was based on the supposition that different mental and emotional functions resided in disparate parts, or 'organs' of the brain. The prominence of the organs reflected the development in an individual of the faculties with which they were associated, and they revealed themselves in the protuberances that make up the surface of the cranium. Gall's student and associate, Johann Spurzheim, brought the teaching to Britain and a Scottish lawyer, George Combe, became its most vigorous proselytiser. Many intellectuals, among them George Henry Lewes, philosopher, naturalist and husband of George Eliot, embraced the new science enthusiastically. Medical opinion was sharply divided, but many of the leaders of the profession became keen supporters. Among these were the most celebrated surgeons of the day, John Abernethy and Sir Astley Cooper, Professor of Anatomy at the Royal College of Surgeons, Vice-President of the Royal Society, remembered now for a pioneering operation to repair aortic aneurysms. The journal of the medical establishment, the *Weekly Medical-Chirurgical Philosophical Magazine*, endorsed phrenology in a series of weighty articles, showing maps of the cranium, in which were defined 'the organ of tenacity of life', 'the organ of self-preservation', followed by organs of choice of nourishment, of the external senses, sexual gratification, reciprocal love of parents and children, attachment and friendship, courage, instinct to assassination (most useful to the forces of the law), cunning, circumspection, instinct of rising in rank or estimation, love of glory, love of truth, sense of locality, sense for collecting and retaining facts, painting and perception of colours, verbal memory, the disposition to learning languages, memory for distinguishing and recollecting persons, liberality, genius for comparison, metaphysical genius, the spirit of observation, wit, goodness, music and theatrical talents, and finally the organs of holiness and of perseverance. And there was besides one 'unknown organ' (but only one).

Phrenology acquired wide popular following and maintained its academic respectability despite what must have seemed to any analytically inclined observer a remarkable dearth of evidence. It was William Hazlitt who wrote in 1829 of the 'thousand instances on record in which this science has been contradicted by facts'. As the nineteenth century progressed, the gospel of phrenology spread and took hold especially in the United States, where 'phrenology parlours' appeared. The psychologist Herbert Spencer, and a leading academic scientist, Benjamin Silliman of Yale University, were numbered among its adherents. Phrenology was proclaimed by the defenders of slavery to demonstrate the inferior qualities of the black (and American Indian) races. Craniology was put to similar uses: the dimensions, and especially the size, of the brain were related to intelligence and other qualities. Paul Broca (p. 10) gave impetus to craniology in France, and throughout the nineteenth century autopsies were carried out on the famous and infamous to report the size of the brain. Reports in the scientific literature—in *Nature* for example—gave the volumes of the brains of recently dead intellectuals and others. When the revered statesman Léon Gambetta died, for instance, it was found that his brain was smaller than average, but to make up for this dismaying finding, its structure was found to be 'very fine, and the third convolution, which M. Broca associates with the speechifying faculty, to be remarkably developed'. Interest in craniology and phrenology have persisted here and there to this day.

An altogether less comical instance of a 'pathological' episode in clinical medicine is presented by the tragic story of the frontal lobotomy operation. This began at the end of the nineteenth century, with the first observation of personality changes, resulting from injury to the frontal lobes of the brain from accidents or the curiosity of doctors in insane asylums. At the same time, others were pursuing the mediaeval idea that violent shocks could reverse mental degeneration, such as the ravages of tertiary syphilis. Precipitation of the patient into ice-cold water from a boat was at one stage a favoured therapeutic measure. Later, the Viennese doctor Julius Wagner-Jauregg treated syphilitics by infecting them with malaria, in the expectation that the resulting fever would effect the desired shock, and might also kill the spirochaetes in the bloodstream. This was later tried in the United States, with dubious results.

Some twenty years on, a veteran and distinguished Portuguese diplomat and brain surgeon, Antonio Egas Moniz, witnessed an operation to remove the frontal lobes of two monkeys, with apparently no ill-effects. Indeed the animals became placid, and the neurotic behaviour that develops under stressful conditions in captivity vanished. Moniz resolved to try the operation on a seriously disturbed patient in a Lisbon asylum. The first victim, a woman, was operated on by Moniz's colleague, Almeida Lima. The plan was not to extirpate the frontal lobes, but rather to destroy the white matter connecting them to other areas of the brain—the leucotomy operation. This was done by drilling holes in the skull and injecting alcohol into the designated region. The patient lived and the operation was repeated on three more inmates. All showed similar personality changes: the agitated and violent behaviour patterns were alleviated, and the patients became vacant and sluggish, and physical side-effects also developed.

Moniz was not dismayed and wrote highly coloured accounts of the success of the treatment in the medical literature. He was applauded by much of the profession, and in 1949 was rewarded with the Nobel Prize for Medicine. In the United States a neurologist, Walter Freeman, and a surgeon, James Watts, took up Moniz's operation and soon introduced a number of technical improvements. In their new version, first tried in 1936, the frontal lobes were totally severed through holes in the sides of the skull. Subsequently the procedure was simplified: an incision was made under the upper eyelid and a blade was slid into the brain. The operation was cheap, and difficult patients became tractable and undemanding, which appealed to asylum directors. It was found that another effect of lobotomy was to control pain, and the operation came to be used for that purpose also: patients with persistent pain reported that afterwards they still felt but were not troubled by the pain.

Moniz's Nobel Prize citation stated that leucotomy 'despite certain limitations of the operative method, must be considered one of the most important discoveries ever made in psychotic therapy, because through its use a great number of suffering people and total invalids have recovered and been socially rehabilitated'—this despite 'sometimes also affecting the intellect, especially highly integrated intellectual functions, such as power of judgement, social adaptability and the like'. The Nobel committee's enthusiasm was premature. In the United States alone at least 50 000 lobotomy operations were performed, until in the early 1960s highly unfavourable follow-up reports, a changed attitude towards mental illness, and the successes of psychopharmacology (which sharply reduced the population in mental hospitals) combined to bring such procedures to a rather abrupt halt.

Notes

1 Dean Swift made fun of it almost two centuries earlier in *Gulliver's Travels*, when he had the Houyhnhnms 'take in at the orifice above a medicine, equally annoying and disgustful to the bowels, which, relaxing the belly, drives down all before it; and this they call a purge'. An enema might also be prescribed for expelling devils; when the nuns in the convent at Loudun in France during the Thirty Years War were thought to have become possessed, the exorcist ordered that the Mother Superior be administered nine litres of holy water in enema. Enemas containing powdered holy relics were also thought by the pious to have miraculous curative properties.

2 The Hippocratic injunction, first do no harm, has even now not taken root in all corners of the healing profession. In the United States a congressional committee found that in one year (1974) 2.4 million redundant operations (known to cynics as 'remunerectomies') had been performed, at a cost of some $4 billion and 12 000 lives.

CHAPTER 8

Science, Chauvinism and Bigotry

There is no national science, Anton Chekhov said, any more than there are national multiplication tables, for national science is no longer science. Today the baleful incursions of chauvinism that Chekhov had in mind seem to have abated, or to have yielded to ideologies of a different stripe, racial, religious or feminist. Fringe movements (which have done more damage in the humanities than in science) have sought to diminish the achievements of the past and the science of today by denying it any validity outside the confines of a white, Caucasian and predominantly male hierarchy. The proponents of these views have not gone so far as to suggest explicitly that the speed of light would be slower, or the number of bases per turn of a DNA helix greater, had the measurements been made by Africans, say, or by women, but their dialectic merges into the 'cognitive relativism', promulgated by a fashionable school of (mainly French) philosophers. It takes as an axiom that, so far from progressing by successive approximations towards a better understanding of the world, the results of scientific inquiry are nothing more than artificial constructs, wholly dependent on the social milieu. This doctrine emerges from a seemingly wilful ignorance of science, verging on obscurantism, and it is strongly reminiscent, as will be seen, of the style of discourse favoured by leaders of totalitarian societies, in their insistence that science must reflect the tenets of the prevailing political system.

That there is no national science does not mean that scientists may not function within a recognisable intellectual tradition and that distinctive styles of working may not make their own unique contributions to the evolution of a field of research. But when tradition is elevated into dogma, science becomes distorted. Chauvinistic pride and animosity were rife during the nineteenth century, especially between France and Germany. Intellectuals in both countries sought to rationalise their assumptions of superiority, even before the hatreds engendered by the Franco-Prussian war. Mutual suspicion and contempt frequently dominated the interactions between French, German, and somewhat less often British, scientists. An illustration is the animosity that developed between the two great panjandrums of German and French chemistry, Justus von Liebig and Jean-Baptiste Dumas, in the mid-nineteenth century. It is hard to be sure what so inflamed Liebig against Dumas when he made his accusations of charlatanism; Dumas, he wrote in a letter to another of the founders of modern chemistry, Berzelius in Sweden, was a tightrope dancer, a Jesuit, a thief—like nearly all Frenchmen. Happily the two adversaries eventually recollected themselves and Liebig visited Paris to receive the Legion of Honour at the hands of his rival.

Nationalistic *hauteur* was endemic, however. Here is the French Minister of Education, affirming in 1852 the superiority that the possession of his incomparable language confers on the Gallic *savant*:

> Does not our tongue appear especially suited to the culture of the sciences? Its clarity, its sincerity, its lively and at the same time logical turn, which shifts ever so rapidly between the realm of thought and that of feeling—is it not destined to be not merely [scientists'] most natural instrument but also their most valuable guide?...Should we, then, having a language with the power to describe with exquisite purity the form of things, deny it the subjects to which it can most usefully apply its admirable precision?

There is much more in the same florid vein.

After the catastrophic defeat of France in 1871, nationalistic outbursts became ever more frequent and shrill in tone. That year Louis Pasteur, who in 1869 had tried, despite having suffered a stroke not long before, to enlist in the Garde Nationale, renounced his honorary degree from the University of Bonn and rejoiced that the conquest of rabies was a French achievement, and that the first victim cured of it hailed from Alsace (which had of course just been lost to the Prussians). Pasteur's letter, which accompanied the rejected scroll, informed the Dean that 'the sight of this parchment is today hateful to me, and I feel myself offended when I see my name, with the description *virum clarissimum*, with which you have decorated it, placed beneath a name destined henceforth to be execrated by my country, that of *Rex Gulielmus*'. The University of Bonn, it should be added, responded with even more flamboyant scorn. 'Sir', the letter (in French) ran, 'The undersigned, acting Dean of the Faculty of Medicine of the University of Bonn, was charged with responding to the insult which you have dared to offer the German nation in the sacred person of its august emperor, King William of Prussia, and to return you the expression of total contempt'. It concludes with a postscript: 'To protect its archives from pollution the Faculty herewith returns your libellous letter'.

Adolphe Wurtz, a distinguished organic chemist and an Alsatian, was another ardent nationalist. In 1874 there appeared his Dictionary of Pure and Applied Chemistry, in which he makes the following ringing declaration: '*La chimie est une science française. Elle fut constituée par Lavoisier d'immortelle mémoire*'. (This makes an interesting contrast to an earlier claim by a German, Lorenz Crell, that Germany was the 'fatherland of chemistry', by virtue of the special mental attributes of its people.) The discoveries during the eighteenth century of Lavoisier—despatched to the guillotine with the remark that '*la République n'a pas besoin des savants*'—were cited time and again as proof that chemistry was indeed a French science. For Lavoisier had annihilated the phlogiston theory, which had dominated chemistry, and vanquished its last and most formidable upholder, Joseph Priestley. Phlogiston (a hypothetical energy fluid, released as flame when substances were burned) was, however, in the view of the French school, a typically Germanic conceit (advanced originally by the German chemist, Stahl), a kind of animism, in

Wurtz's words, 'one of the forms of that nebulous mysticism in which the German imagination finds refuge: the reform accomplished by Lavoisier has been hailed as the retaliation of French reason, the spirit of clarity'. German scientists, Wurtz went on, had forgotten the necessity for common sense in their hypotheses, Descartes's '*bon sens*', which alone tethers the imagination to physical reality. Pascal had expressed the same precept when he wrote that principles were intuitive, while propositions were deduced.

If, a century on, German chemistry was in the ascendant, Wurtz thought, it was only because of a capacity for exploiting the original ideas of others and more especially the organisational efficiency on which the Germans placed such value. German science was run on militaristic principles, with a rigid chain of command, that allowed the

Antoine Lavoisier. Science Photo Library.

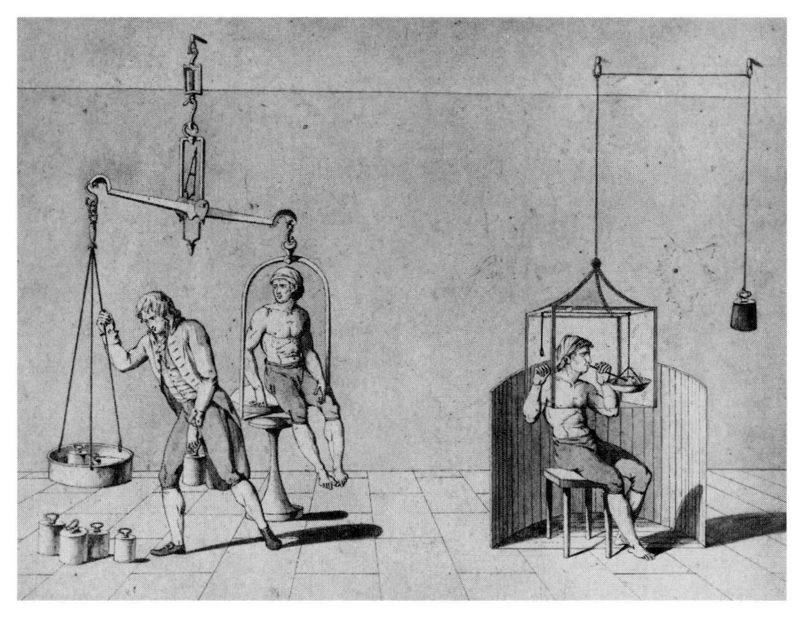

Drawings of Lavoisier's experiments on respiration. ScienceMuseum/Science & Society Picture Library.

Adolphe Wurtz. Science Photo Library.

professor to deploy his army of subordinates to best effect. Wurtz in fact suggested that a little of the same systematic approach (besides a greater investment of resources) would do French chemistry no harm.

The Germans, naturally, had their own ideas. Towards the end of the eighteenth century, the Prussian scientist, writer and aphorist, Georg Lichtenberg, while conceding that the phlogiston theory had been proved wrong, commented: 'That the anti-phlogiston doctrine was initially received with doubt, if not a certain degree of derision, is partly the fault of the nation from which it came. France is not the land from which we Germans are accustomed to receive enduring scientific principles'. This asperity persisted for the next century and more. Hermann Kolbe was an organic chemist of note, a notoriously irascible man, who inveighed against structural chemistry—determ-

Hermann Kolbe. AKG London.

ination of the form of chemical structures in three dimensions—which, led by such men as van't Hoff in Holland and Kolbe's compatriot, Kekulé, was changing the face of the subject. The disposition of the atoms of a compound in space, Kolbe declared, was, and would forever remain, inaccessible to science, and even to discuss it was foolish. Kolbe, a Rhinelander who never set foot outside Germany and was suspicious even of the Prussians, was enraged by Wurtz's claim to the ownership of chemistry on behalf of the French nation. He published in 1870 in his own journal, the *Journal für praktische Chemie*, an article under the title, 'On the state of chemistry in France', in which he delivered himself of some trenchant opinions: French chemistry was of the feeblest, not one university in France could compare with even the least of the German universities, and so on. In his private correspondence Kolbe could not contain his impatience at the forbearance of the Prussians in besieging, rather than overwhelming Paris, and gave vent to his loathing of all things French: 'I have in me deep hatred and contempt for the French, but I have never considered them so uncivilised, vile and beastly, as we now recognise them to be'. Kolbe was infuriated that the names of the cities of Strasbourg, Mulhouse and Metz were not expunged from the cover of the published proceedings of the *Académie des Sciences*, the *Comptes rendus*, and when, some time later, the learned societies of France and Germany sought a reconciliation, he resigned from the German Chemical Society in disgust.

The French physicists, for their part, denounced the obscurity of German physics, with its emphasis, as they saw it, on abstruse theorising. In a widely applauded article, Edmond Bouty and Henri Pellat urged that it should be the mission of the French to lead the subject out of the morass into which the Germans and their epigones in other countries had dragged it. 'Is it not the role of the French spirit', they demanded, 'to shed some light, and bring about the disappearance of all these perilous hypotheses, which will evaporate when one takes pains to examine and quantify them?' But a fully developed hypothesis of national intellectual types was to emerge only during the First World War. It was developed by the illustrious physicist and polymath, Pierre Duhem.

Duhem was never an establishment figure in French science. He taught at provincial universities, finally in Bordeaux. He was something of a nationalist, a devout Catholic and a formidable intellectual, with wide interests and a penchant for water-colour painting, in which he was highly accomplished. He made incisive contributions to thermodynamics (on which he left his imprint in the form of a fundamental relation, known as the Gibbs–Duhem equation), to electricity and to fluid dynamics. In 1915 Duhem delivered a series of four lectures to the Association of Catholic Students of the University of Bordeaux. They attracted large audiences, were reported in the national press, and published the following year under the title of 'German Science', together with an article that Duhem had written earlier in the influential literary periodical, the *Revue des Deux Mondes.*

Duhem was by no means a bigot, and his observations on the Germans and their science were not charged with the incontinent hatred that the war had ignited in other academic circles. Only occasionally does he indulge the audience with what they probably wanted to hear: 'The Germans have continued to boast about themselves. But their tone has changed from one of humility and modesty to one characterised by arrogance, presumptuousness, inordinateness. They used to murmur sweetly: *Deutschland über Alles*! The proclamation of their favourite axiom has now become the furious howling of a pack of mad wolves.' Duhem's thesis is derived from the proposition of Pascal, that there are two types of minds. The French, Duhem instructs his students, pre-eminently possess the *esprit de finesse*, while the Teutonic mind can encompass only the *esprit géométrique.* The first conceives, the second only elaborates. The German mind has its virtues, to be sure, but the concession is grudging: 'Highly skilled at deduction, the German mind is poorly endowed with common sense. It has a limitless confidence in the discursive method, whereas its confused intuition gives it only a weak assurance of the truth. It is consequently peculiarly vulnerable to slipping into scepticism'. Moreover (and alas), 'the mathematical mind of the Germans, so fitted to deduce all the consequences from a given principle, is marvelously adapted to extract an industry of extraordinary power from our mechanics, our physics and our chemistry, the moment they reached a deductive and mathematical stage'.

Duhem then dilates on the authoritarian nature of German learning. The Herr Professor's army of students in a laboratory as large as a factory are subject to military discipline. Each doctoral candidate is expected to examine one inference from the favoured theory, which 'is always verified, without complications, without incident, in the allotted time'. How different from life in the French laboratory! Here 'theories do not always display such docile complaisance'; for the French *savant* knows that, however complete and well-documented they may be, 'they endlessly reveal themselves as over-simple...Never can the sagacious observer pursue the testing of a theory for a considerable time without discovering unforseen, difficult or exceptional cases, upon which his subtlety of spirit finds many occasions to exercise itself. By contrast, at the threshold of certain German laboratories one could write, as on certain lotteries in fairs: Here everyone wins.'

The German, Duhem declares, warming to his theme, 'is demented. His reason is a monstrous thing, wherein an excessive development of one faculty has aborted the other. Endowed with a powerful geometrical intellect which allows him to deduce with extreme rigour, he is deprived of common sense, of that subtlety of intellect which supplies the intuitive knowledge of the truth'. So German science has come to resemble mediaeval scholasticism in its dogmatic adherence to received theory. Its only virtue is a highly developed capacity for building on what its betters have conceived. It stands, says Duhem, in its relation to Gallic science, as the mason to the architect. In short, '*scientia germanica ancilla scientiae gallicae*'.

Duhem was especially distressed by Einstein's Special Theory of Relativity, which had burst upon the shores of physics in the years before the war. He saw it as an affront to common sense and an attempt to undermine the very structure of physics. He was outraged by the tenet that the velocity of a body could never exceed that of light: 'For a supporter of the principle of relativity to speak of a velocity greater than that of light is to pronounce words bereft of sense... That the principle of relativity disconcerts all the intuitions of common sense does not excite the distrust of German physicists. Quite the contrary, to accept it is, by that very fact, to throw over all the theories that speak of space, time and motion, all the theories of mechanics and of physics. Such a devastation possesses nothing displeasing to German thought'. And so on, with ever-mounting outrage. The striking feature of this philippic is the exactness with which it prefigures the anti-Semitic diatribes from the advocates of *Deutsche Physik* during the Third Reich; these zealots saw in the new physics of relativity and quantum mechanics, which had been so displeasing to Duhem, a creeping pollution of their pure, pellucid German science by a miasma of alien, Jewish thought (Chapter 10). On much the same grounds the incursion of predominantly Jewish counter-intuitive theorising into classical physics was denounced by the ideologues of Stalin's Soviet Union. To Duhem, the Catholic intellectual, the irony would have been insupportable.

It was the Great War that drew the most intemperate chauvinistic passions from the European men of science. Few of the leading intellectuals on either side were guiltless. In Britain, citizens with German names were reviled and insulted. They included such essentially English figures as the physicist, Sir Arthur Schuster, whose son was badly wounded on the Western Front, and in the public arena, Lord Louis Battenberg, the First Sea Lord, who was forced to resign and shortly after changed the family name to Mountbatten. Dachshunds were kicked in the street and German-sounding expressions were expunged from the language. (When the United States entered the war, the rather extreme measure was proposed to rename German measles, sauerkraut and hamburger steaks liberty measles, liberty cabbage and liberty steaks, but it does not appear that the expressions ever caught on.) Among the relatively few prominent academics to take a stand against the mounting hysteria were Albert Einstein in Germany, who denounced the war and was in turn denounced (and continued to attract obloquy for his pacifism in later years), and in England Bertrand Russell, who went to prison as a conscientious objector. Some voices were raised in scientific journals, such as *Nature*, entreating colleagues to uphold the international unity of science, which transcended politics, and not to return honours to universities and academies of the enemy countries. Had not Sir Humphry Davy travelled unmolested through France during the Napoleonic wars, visiting his friends and declaring that science knew no national enmities and recognised no frontiers?

To scientists in the countries of the Entente the last straw was the infamous proclamation, known as the Fulda manifesto, or 'The Appeal of the Ninety-three Intellectuals'. Entitled *An die Kulturwelt! Ein Aufruf*, it appeared on 4 October 1914 in all German newspapers and was translated into ten languages. It declared that the signatories—leading figures in the arts and sciences, including among their number a large proportion of the most famous names in science—were united with the German army and rejected 'before the whole civilised world the calumnies and lies with which our enemies are seeking to besmirch Germany's undefiled cause…*It is not true* that Germany is guilty of having caused this war…that we encroached on neutral Belgium…*It is not true* that the combat against our so-called militarism is not a combat against our civilisation…Were it not for German militarism, German civilisation would long ago have been destroyed…The German army and the German people are one'. The country of Beethoven and Goethe knew how to respect the European cultural heritage and its citizens could not have been responsible for atrocities and acts of vandalism, such as the destruction of the great library of Louvain. The manifesto was reinforced by two others, the first from three thousand faculty members of German universities, the second by the Rectors of twenty-two of them, all upholding the unassailable virtues of German culture and behaviour.

It is true that several of the signatories, among them two of Europe's greatest scientists, Max Planck and Emil Fischer, came to regret lending their names to the manifesto, and Fischer later drew up a counter-manifesto, which bore the names of several of the original signatories. But the damage was done, and the French especially could not forgive. Attacks on all things German, and, in the scientific journals, all products of German science, proliferated. Every human vice and failing of intellect was imputed to the barbaric Teuton, and every past affront was recalled. After Duhem's fulminations there followed a series of twenty-three vituperative articles by various men of science in *Le Figaro*, also under the heading of 'German Science', and then in 1916 a book by a professor of biology, Pierre-Jean Achalme, with the uncompromising title, *La science civilisée et la science allemande.* The contribution to the *Figaro* by the Professor of Physiology at the Sorbonne, Albert Dastre, encapsulates the general tone: 'For us science, culture, civilisation formed the common heritage of all humanity, transcending conflicts of interests and national disputes, and lying outside the province of war. Our neighbours and enemies did not understand it in this way. They annexed the universal virtues: beside these venerable words they wrote their—the German—possessive; they spoke of "German culture", "German science", as defining an arm of the German state...[the ideals] reduced to profit, usefulness to Germany, to its supremacy, to its domination'. And again, like Duhem, Dastre insists that the Germans were never initiators, could never aspire to intuitive genius or even appreciate it when they saw it, which was why they had visited on humanity 'a scandal and aberration of the intellect'.

Other of the *Figaro* articles denounced the German approach to medicine, as driven by the innate brutality, the cult of violence inherent in the psyche of the tribe. A philologist enumerated 'the ruse that deceives, the perfidy that betrays, the cruelty that terrorises, the cult of material force, the lie, it is these that characterise the German through the long succession of the centuries'. Some French academics, it is true, took exception to the tone and more especially the assertions of national differences in science, and a Professor of Medicine contented himself with the observation that the Ninety-three Intellectuals had betrayed their calling: they had misused their professional prestige to give weight to a political pronouncement, which they were no more qualified to make than anyone else (an enduring criticism, equally applicable to many publicly conspicuous scientists to this day). These voices of relative moderation, though, were few.

In Britain the assault was led by the chemist Sir William Ramsay, Professor at University College London and winner of the Nobel Prize for his discovery of the noble gases. Ramsay had been a leading internationalist and Germanophile. He was fluent in German and had sent his son to Germany to study. Ramsay was a patron, among others, of Otto Hahn, who spent a period in London before the war and was to become famous for discovering nuclear fission. Now Ramsay proceeded to expostulate in the newspapers and scientific journals against the iniquities of the Germans through the ages. Here he is, writing in *Nature* in 1914:

William Ramsay. ScienceMuseum/Science & Society Picture Library.

> The nation, in the elegant words of one of its distinguished representatives, must be 'bled white'. Will the progress of science be thereby retarded? I think not. The greatest advances in scientific thought have not been made by members of the German race; nor have the earlier applications of science had Germany for their origin. So far as we can see at present, the restriction of the Teutons will relieve the world of a deluge of mediocrity. Much of their previous reputation has been due to Hebrews resident among them; and we may safely trust that race to persist in vitality and intellectual activity.

(He was evidently unaware of the intense patriotism of most assimilated German Jews, who threw themselves eagerly into the war.)

Ramsay, like his French brethren, did not stop at science. He enlarged also on past historical lapses and on such matters as German trade practices, which he described as uniformly unprincipled and predatory. The reason for the war, he opined, was 'the existence in the Teutonic mind of an ideal entirely different from that in the mind of other races'. And again, 'What we have to face is a nation organised for a policy of dishonesty, and a nation, which, as a nation, approves of that policy'. The Germans were also being accused of exalting their own trivial achievements in science and suppressing those of their betters. Here is a typical tirade, published in *Nature*:

> As one of the conditions for their being granted peace by those who shall conquer them on their own selected arena of brute force backed by perverted machinery and prostituted chemistry, they should be made publicly to acknowledge the enormous benefits to science made initially, not by themselves, but by those whom they forced to become their enemies, the Italians, the British and the French…Are the Germans grateful to us for what we have done in science? Do they realise, when they use railroads, steamers, dynamos and telephones, that they are all of British origin? They realise nothing of the kind. Not only are they not grateful for the benefits conferred on them by British science, but they have entered into a conspiracy of silence with regard to them.

And to show that members of the American scientific élite thought no differently, here is an observation by the physicist, Michael Pupin, in a letter to the astronomer, George Hale: 'Science is the highest expression of a civilisation. Allied science is therefore entirely different from Teutonic science'.

When the war ended in 1918, Ramsay was already dead, but many of his compatriots, and certainly many of the French, resolved to exact what retribution they could from the German scientists. There were demands that German chemists who had participated in gas warfare should be tried as war criminals, and highly tendentious victors' accounts appeared in journals, purporting to prove the guilt of the Germans and the innocence of the Allies in this activity. There was fury in France at the award of the Nobel Prize in 1918 to Fritz Haber, who had directed German work on poison gases. Many public figures demanded a full and humiliating repudiation of the Fulda manifesto of 1914 by all its signatories as the price of future recognition by the international scientific community.

The first postwar meeting of the French Association for the Advancement of Science was held symbolically in Strasbourg, reunited at last with the motherland. Also in 1918 the body which represented all national academies of science, the Conseil International de Recherches, voted to exclude the German and Austrian academies, as did the International Commission for Intellectual Cooperation of the League of Nations (the CICI) at its inaugural meeting in 1922. There were calls to eliminate German influence from chemical nomenclature, and many university faculties continued for some years to boycott German scientific journals.

The Germans retaliated as best they could. The professional association of university teachers inveighed against those who wished to resume relations with their *confrères* in the countries of the Entente, and when in 1924 the the members of the CICI relaxed their posture and invited Germany to join, they were rebuffed. In time the passions cooled and by the 1930s—in time for the next cataclysm—the breach had overtly healed, although suspicions remained close to the surface on both sides. It is noteworthy that chauvinistic eruptions did not recur when the Second World War began. The German academic community was no longer unanimous in its desire for war or in its support for National Socialism, and it had in any case already been largely destroyed from within (Chapter 10). Jewish scientists, and others whose political and moral convictions were incompatible with the Nazi régime, were welcomed in Britain and more especially in the United States, and were to give the country that had disowned them reasons for regretting it.

Did the distortions of truth that nationalistic passions imposed on the scientists of Europe and America during the First World War, then, react adversely on their science? It is hard to gauge the extent to which the practice of research was constrained by the curious generalisations, so widely accepted, about racial or national attributes. The war certainly shrank the milieu within which scientists operated. Worst of all, it laid the foundations of the ruinous conflicts that arose within the commonwealth of science in Germany during the black years of the Third Reich.

CHAPTER 9

The Climate of Fear

The tragedy of Soviet genetics

The history of Russian biology in the period between the Revolution and the death of Stalin in 1953 is a woeful chronicle of wanton destruction of both a scholarly discipline and the lives of many of its most respected practitioners. A few charlatans, playing cleverly on the rigid ideology that passed for thought within the ruling caste, succeeded in poisoning all of science throughout the Soviet Union and its Eastern European imperium.

State interference in science in Russia predated the Revolution. There was, for instance, a strong tradition in behavioural physiology, which culminated in the famous school founded by Pavlov. The originator of this strand of thought was probably Ivan Sechenov, who took the extreme position that all actions, conscious or unconscious, could be put down to reflexes. In the minds of the Czarist regime such thoroughgoing materialism was inseparable from subversive radical politics; and thus in St Petersburg in 1866 the sale of Sechenov's book, *Reflexes of the Brain*, was prohibited and he was threatened with prosecution on a charge of corrupting public morals.

Pavlov himself, internationally celebrated for his work on what he termed 'conditioned reflexes'—an expression that entered the vocabulary of dinner-table conversation throughout Europe—was no left-wing ideologue; indeed he was appalled by the social consequences of the Revolution. 'If this is an experiment,' he wrote, 'it is one to which I would not subject even a frog.' Nevertheless, the outcome of his research, with its implicit materialism, was highly congenial to the new Marxist rulers and was absorbed into their orthodoxy. Much of the language stemmed directly from the utterances of Engels and of Lenin about science and its relation to society; Marxism was the science of all sciences, and thus subsumed the entire corpus of natural sciences, pure and applied, within its capacious bounds. So even in 1950, long after Pavlov's death, the President of the Academy of Sciences of the USSR (a physicist) was able in his inaugural address at a plenary meeting to berate Soviet physiologists for 'an erroneous and unwarranted revision of Pavlov's views' and for 'deviating from Pavlov's straightforward materialism'. It was time, he declared, to sound an alarm. 'There can be no doubt that it is only by a return to Pavlov's road that physiology can be most effective, most beneficial to our people and most worthy of the Stalin epoch of the building of Communism'. In similar vein, the President of the Academy of Medical Sciences accused a prominent professor of physiology of 'serious deviations from Pavlov's teaching', and, worse, 'an infatuation with the fashionable reactionary theories of Coghill, Weiss and other foreign authors'. Such a rebuke could herald the midnight knock on the door from the NKVD.

The most deadly attack by far was on the science of genetics and was associated primarily with the infamous name of Trofim Denisovich Lysenko. Lysenko was born in the Ukraine in 1898 into an impoverished peasant family—a considerable career advantage by the time he had grown to manhood. He received only a rudimentary education, but was then sent to a horticultural school. His peasant background next secured him entry, as a member of the so-called red intelligentsia, to a degree course at an agricultural research institute in Kiev. Thence he moved to an agricultural station in Azerbaijan, where he seems to have won good opinions for some work on plant physiology. By 1927 he had attracted the attention of *Pravda* and was well launched into his researches on what he called 'vernalisation'.

Lysenko's philosophy, if such it can be called, derived from the theories of an ageing, unlettered aristocrat, down on his luck, Ivan Vladimirovich Michurin, who had sought to restore his fortunes by taking up fruit growing. In due time he established a nursery with the

T.D. Lysenko. AKG London.

hopeless objective of raising the lush fruits of the Caucasus in the hostile climate of the north. He believed he could achieve this by grafting seedlings from southern varieties onto the hardy northern root stock. Michurin was a vitalist and believed in the limitless adaptability of plants. Any idea that the nature of the plant was determined by heredity was abhorrent to him, and in the years after the Revolution he tried to make his ideas heard in political circles. With extravagant claims—such as the breeding of a hybrid of vegetable marrow and melon—he finally succeeded. The time was ripe, for in the early 1920s the regime had instituted a movement of 'peasant scientists', who were to ply their trade in 'hut laboratories'. So in 1923 the Council of People's Commissars gave Michurin, by then nearly seventy, the means to expand his small enterprise, but they also foisted on him some trained scientists to expedite his experiments. This did not work out well, for when attempts were made to evaluate the outcome, Michurin accused his helpers of distorting the truth. The altercation did Michurin little harm and he became a folk hero, written up approvingly in *Pravda.*

Lysenko got to know Michurin and in the early stages of his career gave him credit for his mind-expanding insights. Lysenko's first claim to public notice came with his supposed demonstration that the farm in Azerbaijan could support winter as well as summer crops of peas. The plants perhaps did grow because it happened to be an exceptionally mild winter. At all events, Lysenko was noticed by *Pravda* and lauded as a 'barefoot professor'. But the real breakthrough came in 1929. That year marked the beginning of Stalin's disastrous policy of collectivisation of the farms, when the traditional ways by which the peasants cultivated their smallholdings were proscribed and grain was confiscated for the collectives. The peasants reacted by killing their livestock and, hastened by severe droughts, famines ensued, which claimed an estimated eight million lives. The scientists in the agricultural faculties and institutes, who had no prescriptions for fulfilling the insane promises made by the Commissar for Agriculture of instant gigantic increases in crop yields, were pilloried as useless theoreticians. The rustic pundits, who claimed to understand the ways of plants, were, on the contrary, eagerly embraced.

Of these Lysenko was the shrewdest and most tenacious, and he possessed the further advantage of a fluent tongue and a hypnotic presence; a portrayal of his personality is to be found in Dudintsev's novel, *Not by Bread Alone.* The loss of the winter wheat crop in the Ukraine had been particularly devastating, and Lysenko began to promote his own ideas for averting such disasters. Whether he knew of the phenomenon, recognised during the previous century, whereby untimely temperature changes could affect the growth cycle of grain crops (without, of course, changing the character of the following year's crop) is not clear; at all events, he got his father to soak a sackful of winter wheat from the family farm in cold water and then bury it in snow until spring. Winter wheat, when sown in the spring, would generally not produce ears, but, Lysenko reported, the conditioned plants did form ears and even gave a better yield of grain than the normal crop. The Ukraine Commissariat for Agriculture reacted with the announcement that this remarkable claim would be rigorously investigated, but in actuality success stories were released to the press before anything had been undertaken. The experts, unmanned by widespread antagonism, reinforced by a number of arrests and even executions, were cautious; indeed the leading plant geneticist in the Soviet Union, Nikolai Vavilov, President of the newly created Lenin Academy of Agricultural Sciences, seems genuinely to have thought that all reports such as Lysenko's should be seriously examined. There was also a general disposition among Marxist scientists to embrace the principle of Lamarckian inheritance (Chapter 2). Speakers at conferences asserted that even the bourgeois savants of the West were abandoning their desperate attempts to prove that genotypes (the genetic make-up of organisms) could not be changed by outside intervention. They called for a new approach to science, based on Marxist principles and *partiinost* (partyhood) and a repudiation of bourgeois, class-ridden genetics. Paul Kammerer's experiments (Chapter 2) were cited in support of Lamarckian inheritance, and Kammerer was even then negotiating a high research position in Moscow.

Lysenko's new technique, which he called vernalisation, brought instant fame and he was translated to the All-Union Plant Breeding Institute in Odessa, as head of a new Department of Vernalisation. His claims now grew increasingly grandiose, as more crops—cotton, to be cultivated in the north, potatoes, and so on—were vernalised and gave promise of an unheard-of abundance. All this was revealed in Lysenko's own new medium, the *Journal of Vernalisation.* The kinds of realistic achievements that soil scientists and plant geneticists (in happier days the pride of Russian science) could offer the increasingly desperate and demanding party officials, charged with stemming the tide of disasters that had overtaken Soviet agriculture, were seen as mere betrayal. Nothing less than miracles were needed and these Lysenko promised. Many crackpots arose during this period and attracted eager interest, but it was Lysenko who had the staying power and the ingenuity to fend off demands for rational evaluation of his claims.

Another who achieved enduring political influence was a soil scientist—an Academician, no less—of seemingly invincible ignorance, but great charisma and political astuteness, with the curious name of V.R. Vilyams (actually Williams, the son of an American engineer who had helped to build the Russian railway system). Vilyams was a Vicar of Bray, who retained positions of power and responsibility under all regimes from that of the last Czar to Stalin's. He was chiefly famous for his grassland system: planting a mixture of leguminous and cereal plants in one plot would lead to symbiotic yield increases of a thousand per cent. There was no scientific basis for this scheme or evidence that it worked, but Vilyams successfully resisted any attempts at evaluation. He issued continuous accusations of sabotage against those he regarded as a threat and secured many arrests and executions. Lysenko and Vilyams evidently hit it off and made common cause in later years.

Lysenko quickly ascended to the status of a folk hero, acclaimed by Stalin himself: 'Bravo, Comrade Lysenko!' he called out from the platform at the conclusion of a speech at an agricultural conference, in which Lysenko proclaimed the victory of vernalisation over the plotting of class enemies, 'so-called scientists', who had opposed him by every dishonest means. Lysenko evidently had a keen sense of drama, and there were many stories of his *coups de théâtre.* It was related, for instance, that when Stalin called a meeting of agronomists to advise how a serious problem of diseased potato crops could be overcome, the scientists spoke of breeding resistant crops. This would take perhaps three years, they said. Stalin was contemptuous of such defeatist prevarication and called on Lysenko, who had been sitting silently in the hall. Lysenko came forward, reached into his pocket and laid before Stalin a handful of shrivelled tubers, picked, he said, on the Agricultural Academy farm. Then, from his other pocket, he drew two large and succulent potatoes—'from my farm'. All his experiments were successful. In 1934 he wrote that 'in five flower-pots, in a corner of a crowded greenhouse' he had generated a new and improved strain of spring wheat, merely by crossing strains with long and short 'vernalisation periods' and short and long 'light periods'. The quality of Lysenko's researches can be gauged from his credo that 'in order to obtain a certain result, you must want to obtain precisely that result; if you want to obtain a certain result, you will obtain it...I need only such people as will obtain the results that I need'. (For details of the exchanges between Lysenko and his adversaries, David Joravsky's definitive work, *The Lysenko Affair*, makes absorbing reading.)

At about this time Lysenko met a self-styled Marxist philosopher, I.I. Prezent. Between them they developed a theory of genetics, an incomprehensible mishmash of misunderstood science and folk wisdom. Prezent would appear at meetings with Lysenko and became his John the Baptist. Together they denounced the inanities of Western genetics. They rejected the chromosomal theory of inheritance and poured scorn on the notion of 'corpuscles of heredity', that is to say genes (even though it had been commended by the influential 'Deborinite' school of philosophers, who occupied themselves with the Marxist dialectic of science, and intially received the blessing of the Central Committee). Vavilov contained himself, apparently believing that serious research in genetics could continue separately from Lysenko's deranged pseudo-science. But then the textbook by Vavilov and the staff of his institute, *The Theoretical Foundation of Plant Breeding*, was condemned in *Pravda* as the work of reactionary biologists; it was Lysenko who had discovered the true laws of plants. This was too much for the leading academic geneticists, several of whom bravely spoke up against Lysenko's nonsense. They included some names of international note—Zhebrak, Dubinin, Serebrovskii—most of whom soon lost their jobs. There had also been at least one analysis in a learned journal of the yields of vernalised crops, which concluded that the process had no significant effect. At a meeting at the Moscow House of Scholars in 1936, Lysenko responded to criticisms and demands for evidence and data. What had any of these cloistered academics done to compare with his feats in giving the motherland millions of tons of grain? They occupied themselves, he snapped contemptuously, with the colour of fruit-flies' eyes. This was a keen thrust, for the great advances in genetics in the past decade had been based in large measure on the work in America by Alfred Sturtevant and Thomas Hunt Morgan on the fruit-fly, *Drosophila melanogaster*. Lysenko had coined a useful shorthand for decadent bourgeois genetics—Morganism–Weismannism (after Morgan and the earlier German zoologist, August Weismann, who had formulated the 'germ-plasm'—in effect the chromosomal—theory of inheritance). Lysenko's rant at the meeting drew a shout of '*Mrakobes!*' from the valiant Serebrovskii, meaning literally demon of darkness or, in the political vocabulary of the time, a term of abuse, signifying obscurantist.

Matters finally came to a head at the fateful conference in Moscow in December of 1936, when Vavilov felt compelled to defend himself against Lysenko and Prezent, who responded with the usual abuse. The atmosphere by then was very menacing, for Stalin's Great Terror had begun. Shortly before, the party representative to the Moscow scientific community had publicly denounced an institute director and distinguished human geneticist, Solomon Levit, for deviant views on genetics, and soon afterwards Levit and two other biologists, Israel Agol and Max Levin, both with impeccable Marxist credentials, were branded 'Menshevising idealists' and arrested. They were never seen again. It was probably no coincidence that all three were Jews. Public denunciations of Levit as a pseudo-scientist, corrupted by the foreign racist and fascist ideology, appeared just before the December conference, and at the meeting itself the head of the Science Division of the Central Committee instructed the participants that there must be no mention of human genetics. All but one obeyed. The exception was an American, H.J. Muller, a leader in the field and a committed communist who in 1933 had accepted—much to the gratification of the regime—a permanent position as Director of the Institute of Medical Genetics in Moscow. Muller concluded his presentation with the remark that it was not his but Lysenko's views that could serve to justify racism and fascism, for if heredity could be determined by training, it followed that the downtrodden masses had indeed become inferior. A speaker from the floor shouted that Muller did not understand what the debate was all about, and there were cries of 'Correct!' from the audience. Muller had in fact already fallen, unknowingly, into the pit of Stalin's displeasure, for he had sent the dictator a copy of his book, *Out of the Night: A Biologist's View of the Future*, which contained incautious allusions to eugenics (see Chapter 11). Stalin had been reading the book before the conference and he had not liked what he read.

The meeting was immediately followed by a wave of arrests, and even the secretary who had translated Muller's textbook of genetics into Russian was arrested, together with several of the institute's scientists, and shot. Muller himself escaped, with Vavilov's help, by joining a transport of International Brigade volunteers, bound for the civil war in Spain. (Aghast at the misfortunes he might have visited on Vavilov and other colleagues, he bravely returned for a brief visit a few months later in a vain effort to mitigate the mischief by a show of solidarity. Muller felt himself fortunate to have eluded 'the closing jaws of Stalinism', but in public he kept his peace, judging that an attack on Lysenko from the West, and from him in particular, would not help the friends he had left behind.)

The students of Moscow University demanded the resignation of professors known, or thought, to be hostile to Lysenko, and some eighty scientists lost their jobs; many were arrested and not a few were shot. Curiously, a number of biologists of exceptional courage, who stood firm and mocked Lysenko's brand of science, were not molested. In 1940, for instance, there appeared in the Proceedings of the Academy of Sciences (*Doklady*) an analysis by the great applied mathematician, A.N. Kolmogorov, of the results of plant-breeding experiments published by two of Lysenko's followers. Their data, he showed, obeyed exactly the statistical rules of Mendelian inheritance, which, so far from disproving, they triumphantly confirmed. As for a paper by Kolman (the man who had denounced Levit), purporting to interpret the same results, it showed only his abject inability to comprehend the principles of statistics. Lysenko himself responded in predictable terms: 'We biologists do not take the slightest interest in mathematical calculations, which confirm the useless statistical formulae of the Mendelists...We do not want to submit to blind chance...We maintain that biological regularities do not resemble mathematical laws'. This was followed by a rejoinder from Kolman (who, as a mathematician himself, must have been stung by the imputation),[1] accusing Kolmogorov of perpetuating the errors of the German mathematicians, von Mises and Mach, who had been damned by Lenin himself.

Geneticists henceforth kept their heads down or changed their field. Some, like the weathercock figure of Boris Zavadovsky (who had earlier upheld the Marxist credentials of 'Morganist' genetics), agilely readjusted their beliefs to become ardent Lamarckians. And so matters rested for a year or two. Lysenko continued intermittently to abuse his now vanquished enemies; at a meeting of the Council of People's Commissars he launched a ferocious attack on the Academy of Sciences and on the unhappy Vavilov, whose institute was obediently condemned by the Presidium of the Academy for not opposing 'a class-hostile position on the theoretical front' and deviating from the precepts laid down by Academician Lysenko. It was not until 1940, however, that Vavilov was arrested, while on a plant-gathering expedition in the Ukraine, accused of 'wrecking' (to which he confessed) and espionage (which he resolutely denied), and left to die of starvation and neglect in prison. Others, including the two most distinguished plant cytogeneticists in the country, Karpachenko and Levitskii, shared Vavilov's fate. Lysenko succeeded to the directorship of Vavilov's institute and dismissed most of the staff in favour of his own vassals. Around the country biologists of all stripes became adherents of Lysenko's Lamarckian doctrines, some no doubt out of a genuine acceptance that the party knew best, but most to keep their jobs. Pavlov's disciple and successor, Orbeli, took advantage of his position as head of the Pavlov Institute and Secretary of the Biological Sciences Division of the Soviet Academy of Sciences, to give shelter to as many geneticists as he could accommodate, to work on what was in effect the genetics of behaviour.

No restraints now remained to curb Lysenko's outlandish theorising. He demanded that genetics be totally banished from the realm, inveighed against the evils of Mendelian inheritance ('the ravings of a monk') and turned also against Darwinian evolution, which he had previously claimed to have clarified. Not content with this, he enunciated a bizarre theory of cooperation to replace natural selection: cooperation prevailed between the individuals of each species to promote beneficial traits. Thus the notion of fertilisation of an egg by a single sperm was a bourgeois error: the more sperm cells combined with the egg the greater the vigour of the offspring. Bourgeois biologists, Lysenko declared, had been neutered by the false doctrine of inter-species competition. By fabricated proofs of this principle they were seeking to justify the class struggle and the oppression of blacks by whites. Plants, he said, could suffer 'socialist re-education', as when winter wheat was turned once and for all into summer wheat. More, organisms of one species could be transmuted into a different species by environmental manipulation: Lysenko could at will transform rye into wheat, wheat into barley. 'Our Michurin biology has proved beyond doubt that vegetable species can be engendered by other vegetable species...Everything depends on the conditions in which these plants develop'. Animals shared these Marxist attributes. So song-birds, Lysenko discovered, gave birth to cuckoos; it had escaped ornithologists that, by Lysenko's law of interaction of species, the song-bird had to pay the price of allowing the cuckoo to get the better of it. It had to change its diet to one of caterpillars and it was this that caused it to hatch cuckoos.

The effect of such vapourings on teaching and research in the biological sciences was catastrophic. Here are two examples of the level to which scholarly discourse had sunk in the universities, taken from Mark Popovsky's book, *Manipulated Science.* A member of the Lenin Agricultural Academy, M.I. Khadzinov, who taught at the Plant Breeding Institute in Leningrad, was summoned before the local party committee and interrogated thus:

'Is Shundenko under your supervision?'

'Yes.'

'Why hasn't he presented his thesis yet?'

'Shundenko is illiterate, he doesn't want to study and he is quite incapable of writing a thesis.'

'But it's your business to see that he gets his degree. If you haven't been able to teach him, you must write the thesis yourself.'

Khadzinov dictated the thesis to his recalcitrant student, who was duly awarded his doctoral degree and appointed academic deputy to the director of the Institute. (Later this same Shundenko obliged the NKVD by preparing the case against Vavilov and became an agent of the security police.)

At a public meeting in the same institution a typical exchange occurred, in the course of which a graduate student accused the teaching faculty: 'You are frightened to death of criticism; it touches you to the quick, doesn't it? Why do [Professors] Rozanov and Vulf try to put the matter in this way? Because they are in favour of Vavilov's theory—a harmful theory which ought to be torn out by the roots, since the working class has coped with its problems without help from the bourgeoisie, and is now in control and achieving results...The whole country knows about the debate between Vavilov and Lysenko, and Vavilov will have to change his tune, because Comrade Stalin has said that the right way is Lysenko's way and not

Vavilov's.' Here another graduate student joined in: 'Lysenko has said it straight out—it's either him or Vavilov, "I may be right or wrong," he says, "but in any case there's no room for both of us." That's putting it fair and square.'

In such an atmosphere, in science as in economics, Gresham's Law begins to operate, and the bad drives out the good. In David Joravsky's vivid words: 'The Lysenkoites had forced political salts into the bowels of Soviet scientists, and some began to void themselves in public, the honourable ones fouling themselves alone, the dishonourable trying to rub it off on others.' A series of buffoons and charlatans sprang from the fertile ground like poisonous toadstools. In the Proceedings of the Soviet Academy of Sciences, the once-proud journal *Doklady*, there sprouted learned papers on how rabbits can transmit tick-bites to their offspring, how strains of chickens can be transformed by injecting into their eggs the white from another strain and how remarkable new hybrid chickens could arise when a single egg was fertilised by several cocks. The head of the agricultural section of the State Planning Committee, who had a laboratory in Lysenko's institute, wrote an article with the conclusion that weeds, which reduced the rye crop, did not stem from seeds carried by the wind or dropped by birds, but were products of transmutation of the rye itself. Sceptics were denounced as idealists, metaphysicians, bourgeois obscurantists and liars, the creatures of the West's reactionary, vapid, scholastic biology.

Brave attempts had certainly been made to loosen Lysenko's grip, and indeed in 1945 Andrei Zhdanov, the Secretary of the Central Committee, had supported a call by the geneticists Dubinin and Zhebrak for a new institute, devoted to real genetics, in the Soviet Academy of Sciences. The plan had foundered amidst a welter of accusations of 'cosmopolitanism' and denunciations of bourgeois geneticists in *Pravda*. Efforts to topple Lysenko were stimulated by the crop failures of 1946, and the severe famine that ensued, aggravated by the devastation of farm land in the war. Lysenko, of course. had no answers, although he had received at Stalin's hands some seeds of a multi-eared wheat, which the dictator urged him to develop as a superior crop. Encouraged by Lysenko's evident failures, the biologists resumed their assault, this time with the support of Zhdanov's son, Yuri. Some meetings were organised at which outspoken rejections of Lysenko's views were heard, and Yuri Zhdanov wrote newspaper articles with coded criticisms of Lysenko. He, feeling threatened, appealed to Stalin for protection against his enemies: he was being persecuted by anti-Michurinist elements, and his best efforts to bring about a renewal of Soviet agriculture were being thwarted. He indicated, moreover, that the multi-eared wheat was a triumphant success and that within a year or two a fivefold increase in yields of grain could be expected.

The appeal did not fall on deaf ears: in 1948 Stalin overrode the electoral process of the Lenin Agricultural Academy and appointed thirty-five of Lysenko's followers as members. Departments of genetics in universities and research institutes were shut down, committees were set up in all major centres to sniff out hints of Morganism–Weismannism and to exact declarations of allegiance to the Michurinist canon from staffs. The Minister of Advanced Education, Kaftanov, reviled the traitors, such as the leading cytologist, Koltsov, for 'man-hating ravings that smell of fascist delirium'. Koltsov was 'the type of wild fanatic whom our contemporary Morganist-Mendelists have for an apostle'. Some three thousand scientists were dismissed from their positions, instructions went out from the ministry to reform all biological and medical sciences on Michurinist lines (whatever this might mean), Yuri Zhdanov made a humiliating retraction and genetics was officially banned. This was the zenith of Lysenko's strange career. His portrait appeared in the foyers of research institutes and his statue in public places, and Lysenko busts were sold in souvenir shops.

The practice of genetics thus ceased throughout the country, except for a few clandestine projects in remote institutions. One such seems to have been led by Nikolai Timoféev-Ressovsky, a fruit-fly geneticist of some note, who had been marooned in Germany at the outbreak of war, and had continued to work there until 1945. The extent of his complicity with the Nazis remains a matter of debate. Against all dictates of prudence he chose to remain in East Berlin, to negotiate the survival of the Institute of Brain Research, of which he was Deputy Director, with the Russian conquerors, and was duly arrested. He served some ten years in Soviet prisons, a part of the time in the kind of institution so powerfully depicted by Solzhenitsyn in *The First Circle*, in which scientists were caged and made to work for the state. So far as his friends were concerned, Timoféev-Ressovsky had been claimed by the Terror, but in actuality he was eventually released and given a laboratory in the Urals, where he resumed his researches. (The story of this hardy survivor is told in lightly fictionalised, though perhaps distorted, form in *The Bison*, by a biochemist turned novelist, Daniil Granin.) The valiant Orbeli was replaced at the Biological Sciences Division of the Academy by Lysenko's supporter, Oparin (a scientist of some standing in his area of research, the origin of life, and an academic powermonger of the most unsavoury kind), who undertook with zeal the task of suppressing Lysenko's opponents. The field was now clear for all manner of Lysenko-inspired pseudo-science to thrive.

One G.M. Boshyan, a student at the Veterinary Institute near Moscow, wrote a treatise with the title *On the Nature of Viruses and Microbes*, which in 1949 appeared in all the bookshops. Microbes, it revealed, were in essence immortal: they could not be killed by chemicals or by boiling, only transformed. They could be made to break up into viruses, and conversely, viruses could agglomerate into bacteria. Microbes could also turn into crystals, and the crystals back into living cells. Director Leonov informed the Ministers of Agriculture and of Health that a great discovery had been made in his Institute, far ahead of anything the Americans had achieved. Ten thousand copies of the book were printed and distributed. Boshyan was rewarded with a doctorate before he had even submitted a thesis and given a large laboratory at the Veterinary Institute. The Director then asked, perhaps for reasons of scientific prudence, that Boshyan's pioneering work should be classified and soon a security post was set up in the corridor. The doctoral thesis, when eventually written, was also classified and disappeared into the confidential archives. At this point the Soviet Academy of Medical Sciences roused itself and set up a committee to study Boshyan's extraordinary claims. For a long time the committee was refused access to Boshyan's laboratory or data, but finally its members were given permission to look down Boshyan's microscope at the miraculous transformations: all they saw was dirt on the slides, and after further investigation Boshyan was stripped of his doctoral degree and expelled from the institute.

The most egregious of all the monstrous and comic figures who sprang up in the laboratories of the USSR was perhaps Olga Lepeshinskaya, whom we have already met in passing (p. 33). She was an old Bolshevik, a friend of Lenin's, and had spent many years in exile in Geneva. She had received no higher education, but rose to scientific prominence in Lysenko's slipstream in the early 1930s. Looking down her microscope she had made the remarkable discovery that cells multiply not only by dividing, but also by breaking up into tiny granules, which develop into quite new and varied types of cells. Cells could be generated from egg yolk and other cell-free materials. Best of all, crystals could be turned into cells when nucleic acid preparations (see p. 64) were added. The whole of the established science of cytology was thus void, and the bourgeois Virchowians had been vanquished. (These were the misguided cohorts of biologists who had espoused the doctrine of the great German nineteenth-century pathologist and cell biologist, Rudolf Virchow [p. 220] and in particular his principle of *omnis cellula a cellula*—that cells could arise only from other cells.)

Lepeshinskaya proceeded to demonstrate to her satisfaction that life could be spontaneously generated. She ground up hydras (microscopic aquatic creatures), filtered the homogenate through gauze, poured the filtrate into a flask, which she stoppered and left for some weeks in a warm place, and lo! the fluid seethed with new and dividing cells. Life had been created! (Lepeshinskaya had in effect repeated the result obtained in the nineteenth century by Pouchet, which, Louis Pasteur showed, were a consequence of contamination with airborne spores) Lepeshinskaya sent her manuscript to Stalin, who called to congratulate her by telephone. Later she was joined by her daughter, Olga Pantaleimova, who achieved similar success with dried and incinerated bird-lime. And when in 1945 news of penicillin reached the Soviet Union, Lepeshinskaya was ready with her own gloss: penicillin gave rise to the *Penicillium* mould, not the other way round. She published a book on her discoveries, which drew a scornful comment in a national journal from a group of cytologists. Lepeshinskaya retaliated with vituperation at two meetings.

Lysenko was impressed by Lepeshinskaya's researches and organised a conference to discuss their implications. Her book was extolled as an expression of the biological ideas of Marx, Engels and Stalin and an example to all of *partiinost* in science. The meeting resolved that new textbooks should be prepared, free of all traces of 'idealist concepts', and that editors of journals should be on their guard against insidious Virchowism. At a conference in 1948, with Lysenko now in the saddle, Lepeshinskaya made an emotional speech. 'What happiness!', she cried. 'At last the dialectical materialists have triumphed, the idealists are paralysed and are being liquidated, as the kulaks [the property-owning peasants] were once liquidated.' The idealists must be exterminated and there must be special vigilance in examining those who came to the penitents' stool to ask forgiveness. She went on to speak highly of herself, as a materialist and innovator, while the leading names in Soviet biology were the very opposite.

The journals were engulfed by a tide of papers based on Lepeshinskaya's precepts. They treated of the transformations of cells and the creation of life from inert material. An oncologist explained that malignant tumours originated from something other than cells. The head of the Department of Histology and Embryology at the University of Rostov announced that living matter was generated when ground-up pearl buttons were injected into animals. Highly placed biological scientists perpetrated similar absurdities, and

Lepeshinskaya's theories also penetrated into the practices of agriculture. At the age of eighty she was still busy, by then at the Institute of Biophysics at the Soviet Academy of Sciences, assisted by her daughter.

It was not until the early 1950s that the wind began to change and Lysenko began to feel the chill. His extravagant promises started to catch up with him. It was becoming all too plain that vernalisation had produced no miracles, not even verifiable improvements, nor had a widely touted programme for crossing dairy cows with beef cattle to produce huge increases in butterfat content of milk yielded any benefits. In addition, Lysenko had spent much of the war in Siberia, overseeing forestry projects. In accordance with his theory of cooperation, he had decreed that oak seedlings should not, as custom and good practice demanded, be planted at widely separated intervals; they should be put down in clusters, so that the weaker brethren would die, sacrificing themselves in the interests of the stronger. The prescription was a disaster, and a few years later nearly all the saplings had died. Stalin evidently had begun to doubt the reliability of his protégé and delivered a speech in the course of which he allowed that science progressed through conflicts of opinions, a principle that had been insufficiently observed of late. This was interpreted as a censure on Lysenko and the opposition grew bolder. There were reports of faked results on transformation of species; biologists hostile to Lysenko were appointed to positions of power and Lysenko's friends, sensing the quarter from which the wind now blew, began to desert him and to disparage his doctrines. In 1953 Stalin died, and Khrushchev, though powerfully biased in Lysenko's favour, did not interfere to any large extent with the new-found autonomy of the Academies.

All the same, it could still be dangerous to dissent from what was even yet widely seen as Marxist orthodoxy. Even in the privileged fastness of Arzamas-16, the new town, far in the south, built to accommodate Stalin's atom bomb project, restraint was prudent. When in 1952 a loyalty commission descended on the physicists, Andrei Sakharov, who in later years was to become the most prominent of the political dissidents, was asked his views on Mendelian genetics and the chromosome theory. He replied that it was, scientifically speaking, correct. The commission showed its displeasure, but recognised that the great physicist was too important to the project to be harrassed. But when Lev Altschuler, another noted theoretician, gave the same answer, he was dismissed from the project. Sakharov then intervened and told the general in charge of security that Altschuler was indispensable to progress on the bomb, and he himself would undertake to curb his 'hooligan conduct'. Altschuler was reprieved but nevertheless ordered to Moscow and given a severe dressing-down by the political director of the bomb programme: he had deviated from the party line in matters of biology, music and literature. 'If everyone was permitted to say what he thought,' the official concluded, 'we would be wiped out, crushed.' Altschuler, though he sinned again, survived, but only just.

In 1955 the Soviet Academy of Sciences appointed one of the country's best biochemists, V.A. Engelhardt, as head of its biology division, and he made it his business to re-establish genetics as a respectable discipline. Even so, as late as 1958 *Pravda*, with Khrushchev's support, called for the editors of an important journal to be sacked for deviating from the Michurinist line. Both *Pravda* and Khrushchev stuck to Lysenko until Khrushchev's deposition in 1964, when the long nightmare of Soviet biology finally ended. A significant defeat was the vote against a candidate nominated by Lysenko for election to membership of the Soviet Academy of Sciences. Engelhardt implied that the man was a nonentity, and Andrei Sakharov followed up with a powerful speech, declaring that those who voted in favour would stand exposed with Lysenko for supporting the infamous and painful episode in Soviet science now happily coming to an end. President Keldysh attempted to stem the applause that followed, but did not allow Lysenko to intervene. In general, voting in the Academy was controlled by the Central Committee itself through the party members, who constituted a majority of the academicians, but in this case the vote went against Lysenko's nominee by 126 to 24.

A committee set up to study the state of Soviet agriculture now reported the incidence of fraudulent claims. Lysenko was removed from the directorship of the Lenin Academy of Agricultural Sciences, but was not tried for his misdemeanours or subjected to the indignities and privations that had been inflicted on his opponents. Instead he was exiled to an experimental farm near Moscow, where he ended his days in tranquil though no doubt resentful seclusion. His death in November 1976 went almost unremarked. Whether he was ever haunted by the ghost of Nikolai Vavilov, posthumously rehabilitated twenty-one years earlier, in August of 1955, or by those of his many other victims, is not revealed.

It remains a matter of conjecture how many biologists—apart from all the 'barefoot professors' and the like, who seized the chance of turning themselves into intellectuals—genuinely believed in the validity of all or some of the preposterous theories of Lysenko and Lepeshinskaya, and how many merely followed the path of professional advancement. The established biologists, who did not protest, probably fell into the second category, but among the younger researchers and students, the political indoctrination to which they had been so relentlessly subjected during their formative years probably stunted their critical faculties. Indeed, it seems often to have been the brighter students who were most vociferous in their demands for a Soviet biology, which rejected the teachings of Western science. Perhaps their intellectual acuity was honed on the problems of reconciling the principles of scientific inquiry with the nebulous and ever-shifting political dialectic of the day. Some members of the emerging generation, accustomed to these mental contortions, quickly adjusted their positions and established themselves after Stalin's death as members of a new élite.

The spread of the contagion

It did not take long for the Lysenkoist contagion to spread to the Soviet satrapies. Some countries were more affected than others, but where it took root, research and teaching in biology withered and died. Worst afflicted was probably Poland. This was also the most tragic case, for Polish science, which had been quite strong before the Second World War, fell victim to German oppression. University professors and lecturers were systematically murdered. Most had joined the army in 1939 and the survivors, who then fell into the hands of the Russians, were done to death in the Katyn forest. When the war ended reconstruction began and scientists and other intellectuals, who had gone underground or passed the war years in exile, returned to rebuild the cultural life of their exhausted country. A student of agronomy at the Jagiellonian University, the country's most venerable seat of learning, in Cracow, has described the dismal scene that met her eye in 1948. Aleksandra Putrament had returned from exile in Russia to find a lack of books and of competent teachers. She estimated that there were no more than a dozen people in Poland with any understanding of genetics. 'Professors from social promotion' were appointed to the faculties on the basis of political merit. The lectures represented biology as a confrontation between the Marxist and bourgeois persuasions and the work of Lysenko, Lepeshinskaya, Vilyams and Boshyan as the genetic orthodoxy.

The only avowed representative of 'Mendelist–Morganist' genetics was Professor Wacław Gajewski, Professor of Botany at Warsaw University, and he has given his own account of what happened in the Polish academies after Lysenko's rout of the geneticists in 1948. Some of his colleagues, he found, willingly embraced Lysenko's 'new biology', others felt it prudent not to demur. Among the unhappiest was Professor Skowron of Cracow, a Western-educated geneticist, whose ill-luck it was to publish his textbook of genetics in 1948, just as the Lysenkoist tide surged into Poland from the East. It was of course the wrong type of genetics, and the hapless Skowron, terrified that he would share the fate of Vavilov and the others in Russia, crept surreptitiously round the bookshops of the city, buying up all the copies he could find.

Professor Gajewski refused the demand by the Council of the Biological Faculty of his university to expunge all 'bourgeois' genetics from his courses, and after a heated debate a compromise was offered: both genetic doctrines might be taught together. Gajewski again refused and was then forbidden to teach at all. A senior member of the professorate, Professor Petrusiewicz, by all accounts a benign and well-intentioned man, thereupon took Gajewski to Russia, so that he might meet Lysenko and be convinced. The visit was not a success. Lysenko began the conversation in characteristic vein: 'If you will not believe what I am going to say, your visit is pointless.' He then embarked on a diatribe that lasted two hours, in the course of which he favoured his guests with many of his insights. Greenhouse plants, he averred, grew only in the greenhouse and not in the open because they had been brought up there; plants do not take up minerals from the soil directly, but through bacteria and fungi; cuckoos do not lay eggs in the nests of other birds, but rather cuckoo chicks hatch from the eggs of the hosts; life could arise by spontaneous generation, and so on. Gajewski departed shaken but resolute.

Professor Dembowski of Warsaw was a noted animal behaviourist, who had worked at the Soviet Academy of Sciences in Moscow and later became President of the Polish Parliament. He had developed a Marxist concept of evolution even before the war, and in 1949 organised the first conference in Poland on Lysenko's biology, only months after the fateful meeting the previous year in Moscow. And Petrusiewicz made public appeals for the suppression of 'idealist' views in science; he formulated a theory of 'creative Darwinism', grounded in dialectical materialism and embodying Lamarckian inheritance of acquired characteristics. The Dean of Biological Sciences at Warsaw University claimed to have shown that chromosomes were a figment, generated by poor technique in microscopic preparation. New textbooks, extolling Lysenko's biology, were published in Poland, and supporters of his philosophy arose in their scores. So matters continued for some eight years, when in 1956 Lysenkoite biology was suddenly abandoned. Three years later the first Western textbook of genetics appeared in Polish translation and in 1960 courses in genetics were introduced into the curriculum as though nothing had happened.

The other countries of the Eastern Bloc do not seem to have suffered as severely. In East Germany one man appears to have been largely responsible for fending off Lysenko's malign influence. This was Hans Stubbe, a plant geneticist, who, although an avowed communist, had survived the political purges during the Third Reich; he had indeed served the regime all too well during the war by organising the plunder of Russian agricultural laboratories and seed collections in the wake of the advancing German armies. The war over, Stubbe returned to his political roots as head of the Cultivated Plant Institute of the DDR Academy of Sciences; there he implemented a research programme in genetics which gained international respect.

In Czechoslovakia there was certainly persecution of politically unreliable intellectuals, many scientists among them, but there were no politically influential proselytisers for the new biology. Miroslav Holub, celebrated as a poet and well-known as an immunologist, has written reminiscences of the period. Party time-servers had risen to occupy most positions of power, in science as in all other spheres. They were, in Holub's words, scientific obscurantists and derelicts. The director of Holub's institute was supposed—though evidently nobody knew for certain—to have been trained as a microbiologist, and it was said that his doctoral thesis had been written by prisoners, who translated its contents from a Bulgarian text. In the Czechoslovak Academy of Sciences, the élite institution, established in 1952 on the Soviet model, some 90 per cent of the scientists were party members. These had been, according to Holub, generally the brightest students, and their loyalty to the party was no sham. They accepted fully the supremacy of Lysenko's new biology—the transformation of species and all the rest. Holub at this time was working in a hospital in Prague as a pathologist, a profession he thought was less inclined than most to renounce 'Virchowian' principles, for Virchow was the principal founder of their discipline. Lysenkoite biologists from the Academy were accordingly sent out to convert the unbelievers, and Holub describes the scene when a young zealot came to lecture to the hospital scientists on his successes, following Lepeshinskaya, in culturing cells from egg yolk. He also told of his heroine's other achievements in the medical sphere, especially the manner in which she had cured herself of pulmonary tuberculosis by bathing in a solution of sodium bicarbonate (baking soda). Too much, thought the assembled pathologists, and one asked a question: what were the full details of the procedure—the concentration of baking soda in the bath-water, the dimensions of the tub and so on? The lecturer was nonplussed: 'I am sorry,' he solemnly rejoined, 'but the volume of Olga Borisovna's bath-tub has not been published.' The audience dissolved in mirth, much to the speaker's bafflement. Holub concludes that the lecturer ended sadly, for his Ph.D. thesis on cells from egg-yolk failed, Lysenko having in the interim fallen from power, so that the research supervisors—one a biochemist, the other an immunologist—blamed each other for devising such an absurd research project. The student became an alcoholic, but the supervisors, whom Holub does not name, became internationally famous scientists. A well-judged change of sides was then, as earlier, the key to survival.

In China Mao, the Great Leader, also expressed his views on Marxist science. According to Engels and Lenin, matter was infinitely divisible: so much for particle physics. The universe, moreover, was infinite in both space and time; so there could have been no point of zero time, no beginning, and the Big Bang therefore never happened, All Western physicists from Newton to Einstein, Mao intoned, were bourgeois; quantum mechanics, relativity and all of cosmology were reactionary. But the Chinese atom bomb was constructed, so his lash could not have fallen too heavily on the physicists.

And what of the communists in the West? The reports of Lysenko's revolutionary achievements in agriculture were initially welcomed, and many biologists with communist leanings, such as J.B.S. Haldane in Britain, were all too reluctant to concede, when the evidence became available, that they had been fooled. At least two books expounding Lysenkoite biology as the new orthodoxy were published in Britain. In France, the communist intellectuals assimilated the 'new biology' into their philosophy. Thus Robert Cohen: 'The victory of Michurinism has put biology on its feet.' Or in the words of a Professor in the Faculty of Medicine in Paris: 'Their thesis springs from the very principles of dialectical materialism, the most powerful tool in scientific thought.' Georges Cogniot, in the Communist Party paper *L'Humanité*, hailed the dawn of the new biology, which was to change the living world. 'Two diametrically opposed tendencies have crystallised in biology,' he wrote, 'the progressive materialist tendency, based on the Soviet naturalist, Michurin, and the idealist and mystical tendency, founded by the reactionary biologists, Weismann, Mendel and Morgan.' A decisive blow had been struck, he continued, barely pausing to wipe the foam from his lips, at the 'theory of mutation, a theory that is the enemy of all rational thought'. Even in 1955, the Marxist theoretician Roger Garaudy could write: 'The alleged disgrace of Michurinist biology in the Soviet Union is nothing but a falsification for political ends.' One can conclude only that the ingestion of too much political dialectic can paralyse the keenest intellect.

Scientific relations between the Soviet Union and the Western countries were in fact exiguous and strained, often paranoid, through most of Stalin's reign and even that of Khrushchev. After the Second World War, however, partly at the prompting of influential scientists such as Pyotr Kapitsa, there was a cautious move to lower the barriers and take note of what was happening in the laboratories of the West. *Doklady* introduced a page of summaries of interesting papers in Western journals and an abstract in English was appended to Russian research papers. Because, in science as in all other activities, the Soviet Union must be seen to outstrip, or at least equal the West, it no longer made sense to draw a line between Marxist and bourgeois science; Soviet achievements, rivalling those of the West, were to be promoted through translations. An International Publishing House was accordingly set up and a number of visits by Soviet scientists to Western Europe and America were permitted.

But paranoia and xenophobia soon reasserted themselves; they were reflected most strikingly in an episode that broke in 1947, described in lively detail by Nikolai Krementsov in his book, *Stalinist Science*. It became known as the 'KR Affair'. The story begins some years before the war, when two researchers in Moscow, Nina Kliueva and her husband, Georgi Roskin, she at the Soviet Academy of Medical Sciences, he a professor at Moscow University, initiated some research on the effects of trypanosomes (the parasitic agents of sleeping sickness) on the growth of cells in mice. The parasite destroys the cells that it enters, and Kliueva and Roskin observed that both live and dead trypanosomes inhibited the growth of implanted tumours. The war intervened and the research was not resumed until the German armies retreated in 1944, when new cultures of the parasites were obtained from Britain through the good offices of the Deputy Head of the Ministry of Health, Vasili Parin.

The new experiments were proclaimed a dazzling success in publications in Soviet medical journals, with 85 per cent disappearance of the mouse tumours. Here was the prospect of a revolution in cancer therapy. The news got into the press and was mentioned in broadcasts, which were picked up in the United States. The Surgeon-General took note and the American ambassador in Moscow was asked to find out more. He visited Kliueva's laboratory, offered American help in the form of chemicals and apparatus, and issued an invitation to the two researchers to visit his country and tell their story. He also approached the Soviet Ministry of Health to propose a Russo-American collaboration for the good of mankind. Molotov, the Foreign Minister, was informed of the proposal and when Kliueva and Roskin, lured by the prospect of fame and lavish funds for future work, petitioned the Central Committee for support, Andrei Zhdanov, the Secretary, secured agreement that a new laboratory was to be built for them. In 1946 the researchers published a series of papers, later summarised in a monograph, *The Biotherapy of Malignant Tumours*, the first copies of which went to the members of the Central Committee.

Plans now went ahead for exchanges between the Russian and American laboratories, and the US Surgeon-General invited Parin to tour the leading hospitals and cancer research centres. All seemed set fair for a collaboration when Parin arrived in the United States, bringing with him copies of the book and papers, as well as trypanosome extracts to be tried out in American laboratories. But three months later, in January of 1947, everything turned sour. Zhdanov summoned Professor Kliueva to the Kremlin and demanded to know why the secret of preparing the therapeutic material had been divulged to the Americans without official permission. Further investigations followed, the luckless Parin was interrogated on his return to Moscow and arrested as an American spy. He and his Minister were dismissed from their posts, while Kliueva and Roskin were hauled before a so-called Honour Court, set up by the Central Committee and apparently directed by Stalin in person. In a three-day trial before a selected audience of more than a thousand, in which Zhdanov himself testified, the two scientists were found guilty of deliberate deception and severely reprimanded.

Harsh decrees now went out, warning against 'slavish servility' to the West; foreign departments in ministries were closed down; all interactions with the West had henceforth to be conducted through the Foreign Ministry, and such scientists as were permitted to travel abroad would have to be accompanied by a security officer. The Soviet Academy of Sciences instructed all journals to cease printing abstracts of papers in English and discussing the contents of foreign publications. More Honour Courts were convened to punish unpatriotic activities. One of these condemned and dismissed the respected geneticist Anton Zhebrak, an officer of the Central Committee's science division and President of the Belorus Academy of Sciences, for slighting Lysenko's work in a scientific article. The doctrine of a shared international science was officially dead. An ugly outcome was the extent to which research laboratories, competing for funds, proceeded to disparage their competitors for 'slavish servility' to Western ideas.

Krementsov concludes in his account of the 'KR Affair' that it was wholly contrived: Zhdanov and Molotov had undoubtedly sanctioned the visits to America and the exchange of research materials. If not the unfortunate Klиueva and Roskin, other unsuspecting scientists would have been chosen as scapegoats. Even if the Cold War was gathering momentum, it is impossible to believe that the resurgence of Lysenko was not a major factor behind the new policy of scientific isolation and the rejection of Western influence. Kliueva and Roskin were not made to suffer materially—had there been a genuine presumption of guilt they would have been arrested—and indeed their new laboratory was built and generously equipped. As to the cancer cure, like many others that sparkled briefly, whether in the East or West, it soon sank from view.

Soviet physics: idealism, pragmatism and the bomb

While the quality of Russian biology was patchy, Russian mid-century physics ranked with the best in the world. The names of Ioffe, Tamm, Landau, Kapitsa, Frenkel, Fock and numerous others resonate through the annals of the subject. Many of these had close relations with their *confrères* in the West, whom they were allowed in the years before the Second World War to visit relatively freely. They had been little incommoded by Stalin's 'bolshevisation' of the Academy of Sciences in 1930, when it was made to surrender its autonomy and allow party members to be imposed on it, as well as admitting engineers and industrialists. (But demands for affirmations of party loyalty and for the admission of selected 'peasant scientists' were quietly forgotten.)[2]

The regime recognised the value of physics, as became clear when Pyotr Kapitsa was kidnapped on a visit to his homeland in 1934. He was an experimental physicist, who had been for some years a distinguished member of Lord Rutherford's Cavendish Laboratory in Cambridge. When he returned to Russia on a family vacation, he was detained on Stalin's orders and installed in a laboratory in Moscow.

From left to right: C.T.R. Wilson, E. Rutherford, P. Langevin, Pyotr Kapitsa and P.M.S. Blackett, in Cambridge, 1919. Science Museum/Science & Society Picture Library.

There was consternation in Britain and America. Several leading physicists, sympathetic to the Soviet Union, wrote letters of protest to Stalin, but the dictator was adamant: Kapitsa would stay. Rutherford did not intercede: he thought that, sorry as he was to lose his friend, Russia's need was greater, and he arranged for Kapitsa's gas liquefaction apparatus to be shipped to Moscow. Kapitsa was a man of fearless resolve and great diplomatic skill, and during the years of the Great Terror used his influence with Stalin to save many physicists from the Gulag and death.

Certain of the party *apparatchiks*, chronically suspicious of intellectuals' loyalty, discovered that modern theoretical physics derived its substance from the German school of Einstein and Max Planck, from relativity and quantum theory, both counter-intuitive in their depiction of the world and therefore almost certainly at odds with dialectical materialism. They were allerted to this danger by renegade physicists, who, like a minority in Western Europe and the United

States, were unable to come to terms with the new concepts. In the Soviet Union the objections could be readily politicised, but curiously enough Engels had applauded Einstein for breaking down barriers between the basic elements of the natural world—mass, energy, space and time—and giving materialist expression to geometry as a part of physics. This must have presented a poser for the Marxist ideologues, who were so violently affronted by the non-classical, counter-intuitive nature of relativity theory. The answer lay in Lenin's reprobation in his *Materialism and Empiriocriticism* of the positivist philosopher and physicist, Ernst Mach, who held that man's understanding of his world was limited by what he could observe and that the causes of phenomena were beyond the mind's grasp. So Einstein could be condemned as a 'Machist' (which, though he respected Mach's science, was far from the reality).

The first such strictures in Russia came from a physicist and veteran party stalwart, by name of Timiriazev. Both he and others, such as A.A. Maximov, a voluble Marxist philosopher–physicist, alluded in approving terms to the effusions by Einstein's most venomous enemy in Germany, and champion of the luminiferous aether (which Einstein had eliminated from physics), Philipp Lenard (whom we shall encounter later). To the 'mechanists', as the revilers of the new physics were called, Newtonian physics, incorporating the aether, sufficed to explain all phenomena from atomic to cosmic, and they strove to have their views proclaimed as Marxist orthodoxy. Ferocious arguments developed about whether the theory of relativity was incompatible with or a triumphant assertion of dialectical materialism. The *Soviet Encyclopaedia* was in no doubt: the aether was required. Both the Uncertainty Principle and the Special Theory of Relativity were often denounced as an affront to Marxism: no Marxist should submit to such defeatist bourgeois laws as the impossibility of defining exactly the position of a particle or travelling at a speed faster than that of light.

Sacerdotal decree moreover, was not to be mocked. Three leading theoretical physicists, Lev Landau, Abatic Bronstein and the noted *farceur* George Gamow (who later escaped to spend most of his working life in America),[3] together with two graduate students, sent a letter to the author of the deplorable article in the *Encyclopaedia*, the political overseer ('red director') of the Physics Institute of Moscow University; in it they declared that they would strive to demonstrate the material existence of the aether, and they asked for Comrade Gessen, the author, to lead them in a similar search for caloric, phlogiston and electric fluid (concepts from eighteenth-century science). The consequences were swift and disagreeable, for totalitarian *apparatchiks* were never noted for their sense of humour. The authors of the letter were taxed with overt rebellion against dialectical materialism and Marxist principles. Landau and Bronstein, who were on the faculty of Moscow University, were brought before a specially appointed court with a jury of workers from the university machine-shop and stripped of their status as university teachers; the two students were deprived of their scholarships and forced to leave the university. Later, when the unregenerate Gamow delivered a popular public lecture on the state of quantum theory and was rash enough to mention Heisenberg's Uncertainty Principle, he was stopped in mid-flight by the local party philosopher, who sent the audience away. His university then warned Gamow never to speak in such terms again. Nevertheless, all these quasi-theological arguments about what dialectical materialism demanded or Engels had decreed made little impact on the progress of research and were ignored in all the major physics schools.

Matters took a turn for the worse when Lysenko, fresh from his triumph at the 1948 conference, felt emboldened to extend his doctrines to the physical sciences. Conferences were organised with the open aim of extirpating foreign bourgeois, 'cosmopolitan' and 'idealist' influences in research and teaching. So Boris Gerasimovich, director of the famous Poltava observatory in Leningrad, was savaged in *Pravda* for his close ties with American colleagues—he had joined with Harvard astronomers in a solar eclipse expedition—and obsequiousness to the West for publishing papers in Western journals.

American protests at such attacks merely inflamed the critics and accelerated the rupture of links to the outside world.

It was their science, as much as political unreliability, that brought some of the best physicists, such as Vladimir Fock, famous for his work on quantum field theory—a rarified subject at the very cutting-edge of the new physics—and the great theoretician and free spirit, Lev Landau, into conflict with the party. Both were saved by the intercession of Kapitsa, who wrote personally to Stalin, testifying to their international reputation, which was bringing renown to the motherland, and vouching for their fundamental loyalty. There was no doubt that Kapitsa took chances with his frequent demands, which, as will be seen, nearly cost him his head.

In 1948 the Minister for Advanced Education, Kaftanov, arranged a conference of physicists to discuss the weaknesses of their subject in the Soviet Union. Kaftanov submitted to the Kremlin a fierce indictment of Soviet physics: it was being widely taught without reference to dialectical materialism; modern conceptual developments were inconsistent with Marxism–Leninism; many physicists were taking up idealist postures; textbooks were subversive because they did not stress dialectical materialist interpretations, and there was far too much discussion of the work of foreign at the expense of Soviet physicists. Much acrimony now developed within the physics community, riven by jealousies between, for instance, the physics institute at the Soviet Academy of Sciences and the Department of Physics at Moscow University. Ioffe and Kapitsa, the two most senior figures, resigned from Kaftanov's committee. The rest fell to squabbling, and the most distinguished theoreticians became targets of abuse. Posthumous opprobrium now fell on the greatly revered figure of Leonid Isaakovich Mandelshtam, who up to his death in 1944 had been the doyen of Soviet theoreticians. Weighty epithets were heaped on his grave: he had been guilty of positivism, conventionalism and operationalism, not to mention reactionary Einsteinism and other arcane sins. A volume of his reprinted lectures on subjects including relativity was ordered to be pulped. The Academy of Sciences, under its chairman, Sergei Vavilov, physicist and brother of Lysenko's most illustrious victim, emerged from the brawl with a modicum of credit. Mandelshtam's protégé, Mikhail Leontovich, was invited to prepare a new, mildly bowdlerised, version of the book. The Academy nevertheless eventually passed a resolution that condemned the widespread servility towards Western science, demanded a restoration of pride of achievement and faith in the inexhaustible power of the Soviet people, and urged that all hints of cosmopolitanism—the Anglo-American weapon of imperialist ideological diversion—be 'mercilessly rooted out'.

The committee had a series of meetings, which resulted in savage condemnations of the foremost physicists such as Ioffe, Kapitsa, Frenkel, Landau and Markov. They were variously guilty of grovelling to the West, propagating open cosmopolitanism, and sowing propaganda for unacceptable Western theories. (The 'Copenhagen model' of Niels Bohr, which embodied the dual nature—part particle, part wave—of fundamental particles, attracted particular opprobrium.) The great nine-volume textbook by Landau and Lifshits, which to this day remains the bible of theoretical physics, was damned for promoting foreign ideology and disregarding Soviet achievements.

Lev Landau had already had problems enough. Before he was thirty, he had become head of the theoretical physics division of the Ukrainian Physico-Technical Institute in Kharkov. Although a dedicated communist and patriot, he had been reckless in his utterances, and his name had come up in the confessions exacted in 1937 from two of his friends, arrested, interrogated and shot for 'counterrevolutionary activities'. Nearly a decade earlier Landau had sought to explain a newly discovered radioactive decay phenomenon, which his first patron, Niels Bohr, had thought violated the principle of conservation of energy. Landau associated himself with this view (which later turned out to be mistaken). But it emerged that Engels had pronounced the Law of Conservation of Energy to be inviolable, and when this was discovered by the local party functionaries, Landau came in for severe censure. Kharkov seems in fact to have been seen

Copenhagen conference, 1930. Front row: Christian Klein, Niels Bohr, Werner Heisenberg, Wolfgang Pauli, George Gamow, Lev Landau (second from right) and Hendrik Kramers. Science Museum/Science & Society Picture Library.

as a centre of subversion, for there were many other arrests, including that of two German physicists, one especially (Friedrich Houtermanns) a considerable name; they were deported to Germany in 1940, as part of an exchange against Soviet prisoners. Many Soviet physicists were arrested, often falsely accused by colleagues; some were shot, others sent to Siberia, where, if they were lucky, they were put to work on technological projects in conditions vividly described by Solzhenitsyn in *The First Circle.*

Landau was noted for his intellectual arrogance and for his inability to tolerate bigotry or stupidity, and his wounding witticisms and tactless practical jokes made him enemies. In 1938 Kapitsa brought him to Moscow to head the theoretical division in his Institute of Physical Problems, but in the same year he was arrested—like his friends in Kharkov, for counter-revolutionary activities—maltreated and gaoled for a year before Kapitsa's appeals to Stalin, guaranteeing that his delinquent friend would mend his ways, succeeded. Landau was redeemed in the eyes of the regime by his contributions to the atom bomb project (of which apparently he greatly disapproved) and became a Hero of Soviet Labour.

It was indeed the atomic bomb that saved Russian physics. Lavrenti Pavlovich Beria, the mephistophelian head of the KGB, was given charge of the programme. Kapitsa, who understood Beria's propensities, pressed his luck too far by requesting Stalin to replace Beria by someone more congenial to the scientists. Beria, deeply inimical from the outset to intellectuals in general and scientists in particular, now asked for Kapitsa's head. This Stalin denied him, for Kapitsa had served the state well by introducing a number of important technological innovations, notably a greatly improved procedure for producing the oxygen needed for metal work and metallurgy; and Stalin evidently had a soft spot for this brave and resolute man. He nevertheless agreed that Kapitsa should be removed from the atomic bomb project, deprived of his laboratory and placed under what amounted to house arrest, in which state he remained until Stalin died.

The Soviet atom bomb, notwithstanding the information about the construction of the American bomb transmitted by the communist informer in the Manhattan Project, Klaus Fuchs, was a formidable achievement; the resources of the war-ravaged economy were much inferior to those the Americans had been able to deploy, but the quality of the scientists was fully the equal of those in Los Alamos. Physics was proving its utility and so, despite continuing criticism about the physicists' failure to emulate their biological brethren and eschew corrupt Western influences, it was in practice allowed to flourish more or less unhindered. But in March 1949 Kaftanov's committee was in fact planning a further meeting, with the aim of obliterating once and for all the bourgeois idealist concepts of quantum mechanics and relativity from Soviet teaching. Beria now had the good sense to listen to Igor Kurchatov, the director of the atomic bomb project, who informed him that, idealist or not, quantum mechanics and relativity were inseparable from the development of the bomb. Beria was evidently taken aback and consulted Stalin, who cancelled the meeting at five days' notice. Stalin reassured Beria, who was still worried about the physicists' ideological unreliability. 'Leave them in peace', Stalin told him, 'We can always shoot them all later'. On 29 August 1949 the bomb was exploded at Semipalatinsk. Kurchatov reported the good news to Stalin, who responded: 'There will be no war'. The physicists were showered with honours and Lysenko's poison was kept out of physics. Yet their tribulations did not end there, for in 1952 there was a savage new upsurge of anti-Semitism, which emboldened the obscurantists to renew their attacks. Mandelshtam was a Jew, as were many of his disciples (and not a few of the leading Soviet theoreticians generally), and they were therefore open to the accusation of 'cosmopolitanism'—of bringing the Jewish ethos into their profession. At a conference in Moscow Leontovich, who had always spoken up fearlessly in defence of his science and of traduced colleagues, turned on the accusers. He denounced their motives, declared the proceedings a farce and strode out of the hall. Despite the support of several distinguished colleagues, the meeting voted to censure Leontovich and reasserted the heterodoxy of Mandelshtam's posthumous *oeuvre.* The affair of "doctors' plot" erupted at about this time: a number of Jewish doctors, accused of the attempted murder by poison of Stalin and other leaders, were rounded up and shot. A new purge of Jews and intellectuals was evidently in preparation, and would have threatened Leontovich (who was not Jewish) and many leading Jewish scientists, but then on the 3rd of March 1953 Stalin fortunately died. The architect of the terror, Beria did not long survive his master's death before he, too, was arrested and shot.

Is there a Marxist chemistry?

Chemistry offers less scope than biology or physics for ideological distortion and it was quite late in the saga of Soviet science that it felt tremors from the Lysenkoite upheavals. The trouble centred on a representation of chemical structures, developed around 1930 by the celebrated American chemist Linus Pauling (now unfortunately remembered for his eccentric campaign, late in life, in favour of vitamin C as a therapy for a range of conditions from the common cold to cancer). Pauling was the outstanding structural chemist of his day and his resonance theory had a profound impact on chemical thinking during the next three decades. In essence it stated that many chemical bonds could be viewed as hybrids of canonical types.[4]

What now happened is most searchingly recounted in David Joravsky's book, *Soviet Marxism and Natural Science 1917–1923*, and see also the article by Istvan Hargittai for a personal reminiscence. The trouble began in the 1930s when an obscure Professor of Chemical Warfare at the Voroshilov Military Academy, Gennadi Chelintsev, put forward his own alternative theory of aromatic compounds, in which benzene, for instance, had alternating positive and negative charges on the carbon atoms. This was a nonsense, but Chelintsev, in the book in which he elaborated his theory, claimed that it supplanted resonance, which he described as a 'mechanistic' hypothesis. This was a coded expression, like 'idealistic', implying that it was philosophically repellent to Marxists. Others then took up the cry, which mounted to a climax at the start of Lysenko's period of ascendency. In 1949 a paper by two chemists appeared with the title: 'On a Machist theory of chemistry and its propagandists'. This condemned the resonance theory as an artificial construct bearing no relation to physical reality, contrived by the Americans for their own convenience. Thus, 'the theory of resonance may serve as an example of the Machist theoretico-perceptional tendencies of bourgeois scientists, which are hostile to the Marxist world-view and lead them to pseudoscientific conclusions on the solution of concrete physical and chemical problems.' Ah, so!

The orthodoxy in structural chemistry was proclaimed to stem from the work of a Russian organic chemist, Butlerov, who had indeed made considerable contributions during the rise of the subject in the nineteenth century. In 1950 Butlerov's primacy was upheld in a report by a committee set up under the aegis of the

Institute of Organic Chemistry of the Soviet Academy of Sciences. The theoretical concepts emerging from the West, it said, were fallacious and reflected the crisis of the capitalist system. But the distinguished committee drew the line at endorsing Chelintsev's ill-founded scheme. It was said that Chelintsev had ambitions to become 'the Lysenko of chemistry', but the chemists, unlike the biologists, were able to act with resolve and deny his competence. The committee also boldly concluded that, while the resonance theory was idealistic and so would not do, quantum mechanics was needed and should not be linked to resonance. A curious feature of the condemnation of resonance is the perception of the leading British communist biologist, J.B.S. Haldane (who was also a considerable biochemist); in his book, *The Marxist Philosophy and the Sciences*, published in 1939, he had, in his intoxication, called the theory of resonance 'a brilliant example of dialectical thinking, of the refusal to admit that two alternatives (two contributing structures) which are put before you are necessarily quite exclusive'. Another irony was that the begetter of the theory, Linus Pauling, had his passport withdrawn by the US State Department at the time of McCarthy's zenith in 1952 (and was prevented from visiting London for a scientific conference) on grounds of communist sympathies.

Worse was to follow, however, for in June 1951 a conference on the state of structural organic chemistry was held in Moscow, attended by 450 scientists, philosophers and hangers-on. A distinguished line-up of chemists and physicists, who had taught the theory of resonance and defended it against its detractors, now recanted. Accusations of slavish adherence to Western bourgeois concepts, of minimising the achievements of Soviet workers and of perpetrating idealism once more flew. Not only was the theory idealistic, it was also 'mechanistic', in that it sought to reduce chemical effects to the mechanics of the electron, which breached the inscrutable principle, laid down by Engels, that 'a higher form of movement' cannot be reduced to 'a lower form of movement'. There was thus a parallel between Pauling's resonance and Mendel's genetics.

Two physical organic chemists, Syrkin and Dyatkina, who had published an influential book, also translated into English, came in for the most intemperate criticism. Dyatkina was asked, for instance, how she could explain her excessive familiarity with the works of foreign authors. Lysenko was held up as an example to those who were laggardly in their loyalty to Marxist principles. A letter was sent to Stalin, before whom the Soviet chemists abased themselves for past failures in understanding the role of theory in chemistry, and especially allowing the foreign concept of resonance to infiltrate their science. The struggle against bourgeois idealist concepts, the letter concluded, was now beginning; the false doctrine had been exposed and would be eliminated from Soviet science.

The proceedings of the conference were published in 1952, and a series of recommendations was made. It included a devious compromise, amounting in effect to endorsement of a sort of diluted resonance theory, couched, however, in terms that related it to the sainted Butlerov's opinions. This caused outrage in some quarters, nowhere more than in the mind of Chelintsev. He recognised the newly contrived theory as a modified version of resonance, he objected to the committee's failure to condemn Pauling and Ingold, and he charged the most prominent chemists of the day with suppressing his own freedom to publish his work. He aired these sentiments at other meetings and attempted to enlist the help of Lysenko's apologist, the philosopher Maksimov. But Maksimov, sensing a change in the wind, had cooled off. Chelintsev, he said, was trying to emulate Lysenko's Morganism–Weismannism with his own Paulingism–Ingoldism, but these distinguished chemists were not reprobates of the same hue. Chelintsev, abandoned even by his few supporters, was humiliatingly defeated. He kept trying, but to everybody's relief the heat had gone out of the debate and with Stalin's death in 1953 it ended, to surface briefly much later in publications by Chelintsev and others; and in 1958, one Butnev reported that he had found differences between bond-lengths in the benzene molecule. But the accuracy that would have been required to establish the conclusion was beyond the capacity of the method that Butnev was using and the work was quickly discredited. In 1961 Pauling at last came to Moscow, where he lectured on his resonance theory to an audience of 1200 scientists.

Notes

1 Ernst Kolman was a Czech, who had begun his sojourn in Russia as a prisoner of war and had afterwards studied at Moscow University, where he had come under the influence of the Deborinite philosophers. He had been a member of the Soviet delegation, led by Nikolai Bukharin (later seen by Stalin as a dangerous rival and executed after a notorious show-trial), to a widely publicised conference on the history of science and technology, organised in London in 1931 by a group of left-leaning British scientists. There Kolman must have impressed the audience with his mindless adulation of Marx, Engels and Lenin, as guides to mathematical truth. He whipped himself up into a lather of enthusiasm, for instance, over 'the hitherto unpublished writings of Marx dealing with mathematics and its history, of which there are more than fifty... [which] are of tremendous importance'. Compare this with the effect of Marx's and Engels's utterances on physics, which shocked the young Mark Azbel, a theoretical physicist, by their ignorance. This, as he relates in his memoirs, *Refusenik*, put him off communism for good. To Kolman's credit it should be said that he at least defended the physics of Einstein and Planck when it came under doctrinaire attack in later years.

2 The Academy also benefited from an eightfold increase in its budget during the three years that followed. During this period the population of university students rose by a factor of about three, while the number of professional scientists increased tenfold.

3 Gamow was chiefly famous for his work in cosmology—he was the originator of the Big Bang theory of the universe—but he was also a remarkable populariser of science. His books, *Mr Tompkins*

Explores the Atom and *Mr Tompkins in Wonderland*, still make entertaining and painlessly educational reading. Among his jokes was the authorship of a classic paper about the formation of the elements: Gamow had taken on a graduate student, Ralph Alpher, who carried out much of this work, and he then asked for the help of the celebrated theoretician, Hans Bethe, so that the paper could appear under the names of Alpher, Bethe and Gamow. Gamow tried unsuccessfully to persuade another of his associates to change his name to Delter.

4 The theory was primarily concerned with organic chemistry, the chemistry of carbon compounds, which include all biological molecules. The carbon atom is tetravalent, that is to say can form bonds to four other atoms. The simplest carbon compound is methane, CH_4, which consists of a carbon atom bonded to four hydrogen atoms, and its structure is written:

```
    H
    |
H — C — H
    |
    H
```

Methane is the first member of a series of compounds, of which the second is ethane, C_2H_6, written:

```
    H   H
    |   |
H — C — C — H
    |   |
    H   H
```

But in the molecule ethylene, C_2H_4, two of the valences, as they are called, form a double-bond between the two carbons, leaving two left over from each to bind hydrogen atoms:

```
H         H
 \       /
  C  =  C
 /       \
H         H
```

The association of the carbon atoms is tighter in ethylene than in ethane, and the separation between the two nuclei is smaller; in fact the 'bond-length' defines the character of the bond—

whether, that is to say, it is a single or a double (or indeed, as in acetylene, C_2H_2, a triple) bond. In certain cases, notably the so-called aromatic compounds, of which the archetype is benzene, the bond lengths are intermediate between the two. Benzene, C_6H_6, is a ring compound, written:

```
          H
          |
   H    //C\    H
    \ C//    \C/
      |       ||
      C\\    /C
    /    \\C/   \
   H       |     H
           H
```

But in actuality the alternating single and double bonds, which seem to be required if the tetravalency of all six carbon atoms is to be satisfied, are a misrepresentation, for all bonds are in fact the same length—somewhere between those of a single and a double bond. This caused chemists difficulties, which Pauling's resonance theory sought to resolve by depicting the bond as a hybrid between double- and single-bond forms, or as 'resonating' between the two types. Now this was a formalistic representation, and Pauling was explicit that the aromatic bond should not be seen as actually flickering between the one and the other, but rather as partaking of the characters of both. The theory was immensely fruitful, and was extended by organic chemists, notably the American, George Wheland, and, in London, Christopher Ingold. Today resonance has been largely replaced by other, more powerful representations.

CHAPTER 10

Science in the Third Reich: Bigotry, Racism and Extinction

Hitler's Thousand-Year Reich endured for twelve years—a mere fifth of the lifetime of the Soviet Union and its Marxist imperium. Science, like all other facets of the country's intellectual life, was maimed by the Nazi regime, but, as in the Soviet Union, a vein of quality endured. Around it grew the rank and malodorous weeds of Aryan physics, biology and, worst of all, medicine. They were propagated not only by impressionable dunces but also by genuine Nazi zealots inside the scientific community and by a host of cynical opportunists, not all of them inferior intellects. As in Russia, physics proved the most resilient of the sciences and in the main rejected the ideological poison. But whereas Stalin was persuaded of the value of physics to the state and was quite prepared to set Marxist principles aside if they stood in the way of technical progress, Hitler's bigotry was impenetrable to reason. 'If science cannot do without Jews,' he told one of the country's eminent physicists, 'then we will have to do without science for a few years.' The consequences are well known.

The roots of fascist biology

Racism, whether in Germany or anywhere else, has never originated in the theories of scientists and it would be absurd to make Ernst Haeckel out to be the father of National Socialism and of the evils that flowed from it. But Haeckel's science served the Nazi regime as a fig-leaf, and gave rise to a hideous pseudo-science, which Haeckel, ferociously nationalistic as he was, would probably have abjured.

Ernst Haeckel lived from 1834 to 1919. He was Darwin's foremost champion in Germany, and while Thomas Henry Huxley was dubbed 'Darwin's bulldog', Haeckel was to his British admirers Darwin's German bulldog. For much of his working life he was immersed in a relentless feud with his sometime patron, the great Rudolf Virchow (who, it will be recalled, occupied a high position in the demonology of the Soviet Lysenkoites). Virchow was the founder of modern pathology, a chilly, formidable and acerbic personality, not noted for self-deprecation, who became known as the pope of German medicine. He was also a political radical, and incurred the disapproval of his government when he reported on the state of famine and disease in Upper Silesia, then an impoverished region, blaming its woes on

Ernst Haeckel. Science Photo Library.

Rudolf Virchow. Science Photo Library.

the government's maladministration. In 1848, the year of the revolutions, Virchow went to the barricades, and when the risings were put down found himself still deeper in official displeasure. He was later elected a parliamentary deputy for the Progressive (radical) Party, of which he was a founder, and proved a thorn in the flesh of the Chancellor, Bismarck, who at one point challenged him to a duel. (Virchow declined, saying that he would fight only with scalpels.)

When he arrived at Würzburg University to study medicine, Haeckel, having read *The Voyage of the Beagle*, was already captivated by Darwin. Among his teachers at Würzburg was Virchow, whose scope and intellectual rigour impressed the extrovert and boisterous young man. For one summer, in 1856, he became Virchow's assistant at the Anatomical Institute. But Haeckel took little pleasure in medicine and proceeded to indulge his passion for natural history. When *The Origin of Species* appeared in Germany, Haeckel was intoxicated by its luminous depths. Virchow (by then established as professor at the medical school in Berlin), on the other hand, was violently hostile to Darwin's teaching. The Virchowian theory of cellular pathology asserted that all diseases stemmed from what we would now call metabolic disturbances—anomalies in the function of cells—not from imbalance of the 'humours', as had hitherto been taught. Any alteration in cellular function would, conversely, lead to disease; thus in essence all mutations were deleterious, and not, as Darwin argued, the random agents of change. In pursuit of further weapons to use against Darwin, Virchow also entered the field of anthropology and averred that a Neanderthal man he had examined was not a member of a distinct species, a precursor of *Homo sapiens*, but a bandy Cossack with rickets.

What most dismayed Virchow was the rise of the trend now called Social Darwinism, of which Haeckel became a prophet. All social behaviour, according to Haeckel, was governed by evolution through 'survival of the fittest'. The State, the neo-Darwinians believed, functioned as a biological organism, subject to the same laws of evolution. They spoke of the anatomy, physiology and metabolic activity of human societies, as though society was self-evidently the macrocosm of the microcosm, which was the individual. Haeckel equated evolution with progress, as an irresistible law of nature. In the struggle for survival, the fittest races would prevail. Moreover, the hereditary purity of the race must be preserved from contamination by alien stock and protected against genetic deterioration. Social Darwinism thus became ominously linked with demagogic socialism and assertive nationalism. Haeckel was soon appointed to a professorial chair at the University of Jena, and, a charismatic and romantic figure, quickly gathered a large following.

Virchow perceived the dangers inherent in Haeckel's doctrine with great clarity. Consider, he demanded, what the theory of evolution represents to the mind of a socialist: the question might appear frivolous, but it was in reality highly charged. 'I hope that the theory of evolution will not bring down upon us all the horrors that a similar theory visited on our neighbour. [He was alluding to the Terror that followed the French Revolution.] Undoubtedly, if the logic of this theory is followed through, there must be uncommonly disagreeable implications, and the fact of its adoption by socialism will I hope not have escaped you. We must be quite clear about this.'

These prophetic words infuriated Haeckel, who denounced them as malicious and defamatory, and it was his utterances, not Virchow's, that caught the popular imagination. Social Darwinism had an instant appeal for the Germans, and Haeckel, though not its only interpreter, was much the most flamboyant and influential. He insisted that the theory of evolution, with its implication of the eternal struggle for existence, held the answers to all social, intellectual, aesthetic and political problems. To reflect the universality of the concept he called his philosophy 'monism'. It came, some decades later, as an epiphany to Adolf Hitler and influenced the precepts set out in *Mein Kampf*.

To this farago was added a mystical belief in the superiority of the Aryan race, as enunciated by the French romantic philosopher Gobineau in his *Essai sur l'inégalité des races humaines*, which had an equal resonance in Germany. (Its German translator was Ludwig Schemann, an intimate of Wagner and a member of his so-called 'Villa Wahnfried circle'.) 'World history is race history', Gobineau asserted. He was persuaded that the decline of the noble stock was at hand, for it was being corroded all the time by interbreeding with lesser races. The theory of the moral and intellectual primacy of the German race was further developed into a mystical pan-Germanism by the English-born German chauvinist, Houston Stewart Chamberlain. From this potent brew there emerged the concept of 'racial hygiene', with its implicit anti-Semitism and demand for eugenic measures (of which more later).

The term *Rassenhygiene* was coined in fact by Alfred Ploetz, a doctor, concerned, like many who followed, at the prospect that the progress of medicine would promote the survival of enfeebled human stock. It was essential, he proclaimed, to preserve the health and genetic purity of the superior Nordic or Germanic race. (Rather surprisingly, he placed the Jews only just below the Aryans in the human hierarchy.) As a forum for debate on the subject, Ploetz founded in 1904 a new learned journal, the *Archiv für Rassen- und Gesellschaftsbiologie* (Archives of Racial and Social Biology), and a Society for Racial Hygiene to go with it.

Ploetz was followed by a series of apostles of the new order: National Socialism appealed to biologists of the Haeckel school, with its insistence on the integration of biological and sociological laws. Ploetz, who had had no academic standing, was given a professorial chair at Munich University in 1936, when he was approaching eighty, on Hitler's personal instructions. One who came under Ploetz's influence was the zoologist, H.F.K. Günther, known to his intimates as *Rassen-Günther*, who in 1922 published an influential treatise, *Rassenkunde des deutschen Volkes*. This is known to have inspired many early proponents of fascism, Houston Stewart Chamberlain included, and not least Adolf Hitler. The issue of racial hierarchies was a special obsession of students of physical anthropology, a discipline strongly represented in the German universities. Practically all revealed themselves, as soon as it became acceptable to do so, as racists of the most unwholesome kidney. Prominent among them were two professors, Eugen Fischer and Otmar von Verschuer, who both became associated with programmes of experimentation on prisoners in concentration camps and, much earlier, in the sterilisation of the so-called *Rheinlandbastarde*; these were the children of German mothers by French colonial soldiers, who had occupied the Rhineland in 1919. Fischer had made his reputation as an anatomist in the study of the results of interbreeding between races, especially in the German colony of South-West Africa (now Namibia). He had published a book in 1913, *The Rehoboth Bastards and Human Bastardisation*, in which he concluded that this inferior breed would perish under the pressures of biological competition. Even the Rockefeller Foundation, which at the time was generous in its support of science outside as well as inside the United States, granted funds to von Verschuer (whose valued assistant was Josef Mengele of infamous memory) for studies on twins and on the effects of poisons on development of germ cells (the cells of sperm and ovum, which

Eugen Fischer. AKG London.

combine on fertilisation and give rise to the new embryo). Fischer and von Verschuer were participants in a meeting with ministers on racial policy a little before the Wannsee conference, at which the 'final solution' was formulated. Fischer was by then likening alien racial elements, the Jews in Germany in particular, to parasites or tumours, which must be excised from the body politic—an analogy developed, as will be seen, by the ethologist Konrad Lorenz.

If there was a single theme that characterised the biology practised by the academic adherents of the Nazi party, it was an emphasis on 'holism' in all its aspects, and there was no shortage of aspirants willing to take up the cry. The approved *völkisch* medical doctrine

Otmar Freiherr von Verschuer.

was that all elements of the body were interconnected; there was thus no room for clinical specialists or their 'mechanistic' doctrines. Conventional academic medicine was to be replaced by herbs, homeopathy, sunshine and fresh air. The foremost prophet of this school was probably Erwin Liek, an elderly surgeon in Danzig (now Gdansk). Liek's views derived apparently from the mediaeval concept of bodily humours. He argued, besides, that pain was a benign secretion of a disorder and an essential part of the healing mechanism: to suppress it impeded recovery. Endurance of pain was, moreover, a prime virtue and one with which the superior races were better provided. More, to be sick showed a lack of moral strength—was indeed immoral. (Had Liek perhaps read Samuel Butler's *Erehwon*?) This provided Liek's followers, if not Liek himself, with a justification for annihilating the sick. The creeping primitivism of Liek and other such wild-eyed visionaries quickly permeated all medical schools, although arguments of scholastic obscurity over points of interpretation continued even within organs of the state, such as the SS.

Party members and sympathisers filled the professorial positions from which Jews and the politically suspect were expelled as soon as Hitler acceded to power in 1933.[1] Hitler himself had spoken of the underlying laws of life, which had to be laid bare, and the pronouncement by the eminent ecologist, August Thienemann, that National Socialism was in essence applied biology, must have cheered the camp-followers. One who accepted Hitler's nebulous challenge was Otto Mangold, Director of the Zoological Institute of Freiburg University; he enlarged on it in a speech and pamphlet with the title, 'The Tasks of Biology in the Third Reich'. Ernst Lehmann, Professor of Botany at Tübingen, was another who embraced the Nazi racial ethic: it was Hitler, he wrote after the seizure of power in 1933, who had perceived the dangers of racial pollution; it would be 'the task of biology to promote the spread of this knowledge and to forge new weapons for the struggle to come'.

Lehmann referred to the 'biopolitical fight' against racial deterioration: the biological will to win must be inculcated in the German people. The individual must subjugate himself to the interests of the *Volk* (a word that had come to carry mystical, romantic overtones: *Du bist nichts, Volk ist Alles* was the mantra); the self-willed individual was a foreign growth, a cancer, inside the body of the *Volk* and forfeited the right to existence. Biologists must educate the people to understand these principles; they must bring to light the biological imperatives to which Hitler had alluded, and they must incidentally also serve the state in material ways, in agriculture, husbandry and nutrition. Around 1936 chairs of *Rassenkunde* (racial science) were established in many universities, and were generally occupied by obscure party members with no credible academic record. Many belonged to organisations such as the *Kampfbund der Artamanen*, a freshly coined word of ancient, apparently Indian, etymology, implying adherence to the mystical creed of *Blut und Boden*, blood and soil.

The abominable Lehmann had already founded a new journal, *Der Biologe*, and a professional organisation, the Association of German Biologists. Lehmann's overtures to the Nazi party, the NSDAP, were well received and his society was gathered into the thriving Association of National Socialist Teachers. The journal appeared monthly and contained articles—almost all relating to racial topics—and political proclamations, news and announcements. There was also a column entitled 'What we want to read and hear', with quotations from scientific journals and newspapers, offset by 'What we do not wish to read and hear'. In an early issue Clemens Schäfer, a zoology professor at Breslau University, had urged the adoption of Germanic terms in biology in place of those with Greek or Latin roots. Symbiosis (*Symbiose*), for instance, was to be replaced by *Genossenschaft*, or companionship, ecology by *Umweltlehre* (whole-world science) and parasite by *Schmarotzer*, a sponger. The scheme does not appear to have caught on.

Disputes about the theory of evolution were frequently ventilated in *Der Biologe*. Among those who rejected it and derogated Haeckel as a vulgar materialist was a panjandrum of the Nazi scientific establishment, Ernst Bergdolt, who had been made a professor at the University of Halle. Notwithstanding, *Der Biologe* celebrated the centenary of Haeckel's birth. Gerhard Heberer, a biologist with an evolutionary theory of the emergence of Nordic man as a new and more highly evolved species, took on Bergdolt and the anti-Darwinians in a heated exchange

Another prominent contributor and member of the editorial board of *Der Biologe* was Konrad Lorenz, who also called for a policy of racial purification, and indeed of euthanasia for the unfit, and became a member of the party's Office of Race Policy. As a behaviourist, he was

.JAHRGANG HEFT 1/2 JANUAR / FEBRUAR 1940

Der Biologe

Monatsschrift des Reichsbundes für Biologie und der Unterabteilung Lebens- und Rassenkunde des NSLB

Schriftwalter: Reg.-Rat Dr. W. Greite,

Bundesleiter des Reichsbundes für Biologie

Leiter der Forschungsstätte für Biologie

in der Forschungs- und Lehrgemeinschaft „Das Ahnenerbe"

INHALT:

J. F. LEHMANNS VERLAG MÜNCHEN BERLIN

Title page of an issue of *Der Biologe.* Note the article by Konrad Lorenz.

concerned about the degeneration of the race through the comforts of urban civilisation, a process which he compared to the domestication of animals. But he also believed in a hierarchy of races, extending from the *vollwärtig*, full-valued, to *minderwertig* or valueless, and the second were a cancer within the body of the first.

Lorenz was an Austrian, who had shown leanings towards fascism from the outset and had made haste to join the Nazi Party immediately after the *Anschluss* (when his country was conjoined with Germany). In 1940 he secured a professorial appointment in Germany, as Director of the Institute of Comparative Psychology at the University of Königsberg, overriding the wishes of the Faculty. Called up for service as an army doctor the following year, he was appointed psychologist to the military district of the *Wartheland*, the provinces of western Poland, which the Germans claimed for their own. Its centre was the city of Posen, known to the Poles as Poznán, and he also received an appointment at the Reich University of Posen (one of only two such established in occupied cities; the other was in Strasbourg). In Poland Lorenz contrived to carry out research on people of mixed German and Polish descent, arriving at the conclusion that Germans and Poles were 'structurally' incompatible. Like most of the other unpleasing protagonists in the dismal story of German biology, Lorenz exculpated himself shamelessly when the war ended and was (like Eugen Fischer) received back into a state of social and academic grace. In 1973 he shared the Nobel Prize for his work on animal behaviour, appearing in the guise of a benign, white-bearded old gentleman, often photographed creeping in his shorts and on all fours through the long grass, followed by a procession of ducklings, which accepted him as their mother. His popular books on animals, especially *King Solomon's Ring*, were enduringly successful and are still widely read. Those German biologists who had published racist textbooks—Lehmann for one—hastened to get the toothpaste back in the tube: new and sanitised editions were quickly brought out and the transition to ideological respectability was for the most part painlessly effected.

Der Biologe and the Association of German Biologists both flourished—more biologists and doctors than any other professional people had become party members—and in 1939 Heinrich Himmler, the Reichsführer SS, decided he could make use of it for his own purposes: he affiliated it to the most monstrous and absurd of all the pseudoscientific growths in the Third Reich, the Ahnenerbe, which deserves a section to itself.

The Ahnenerbe: Himmler the intellectual

Heinrich Himmler was a *petit bourgeois* with intellectual pretensions. He had studied for a diploma in agriculture and took a dilettantish interest in such subjects as astronomy, which he indulged with the aid of his own telescope in his garden. Himmler was a man of severely limited intelligence and had no notion of the nature of scientific inquiry. He believed, in particular, that truth was vouchsafed through the imagination and that the task of science was to gather proof of revealed propositions, to search for the stone that fitted the hole in the mosaic of knowledge: evidence that did not conform to the revealed truth was the outcome of error or incompetence. Himmler detested the academic establishment, but had a well-justified fear of taking on academic specialists on their own ground. He forsook his Roman Catholic faith for the ideology of National Socialism, above all its assumption of Aryan racial superiority, and it was not long before he developed a consuming interest in the history of the Nordic peoples.

With the power and budget that came with his appointment as Reichsführer SS, Himmler felt able in 1935 to fulfil his ambition and set up a research institute dedicated to his kind of science. He called it the 'Ahnenerbe', or ancestral heritage, and it was to concern itself primarily with the spiritual, intellectual and biological history of the race. As the administrative head, or General Secretary, Himmler appointed Wolfram Sievers, a personable thirty-year-old SS man, a member of the Artamanen, son of a church musician and himself musically talented. Several luminaries of the Third Reich influenced the scope of the Ahnenerbe's programme, among them the Reichsbauernführer, leader of the peasantry, Walther Darré, whose interest was in agriculture and the land, and a philologist, Hermann Wirth. Philology in fact engaged a considerable part of the institute's energies from the outset. Himmler would brook no interference in his plans, and he fell out with the early advisers. It was thus a professor at Munich University, Walther Wüst, and not Wirth who was appointed to head the philological section and was named research overseer for the entire organisation.

Himmler was a deeply superstitious man, much taken up with astrology and the occult, which provided him with evidence that he was a reincarnation of the first German king, Heinrich I (whose millennial celebrations took place in 1936). Himmler wanted the Ahnenerbe to bring the natural sciences to bear on the so-called *Geisteswissenschaften*, the 'sciences of the spirit', but he was never able to attract any natural scientists of stature to his organisation. This is perhaps hardly surprising, given his views on science, and especially biology and cosmology. He held, for instance, that the Aryan race had not been fashioned, like the rest of mankind, by Darwinian evolution. There were no monkeys in *his* ancestry: the Aryans were *göttergleich*, or godlike, endowed with preternatural knowledge and wisdom, and they had descended from the heavens, when released from cosmic ice particles in which they had lain inchoate as tiny nuclei, waiting to hatch. The primeval gifts bestowed on the members of this privileged race included a knowledge of electricity, for thunder and lightning were manifestations of an advanced war-machine that the celestial forbears had designed.

The staging post for the Aryan conquest of the world was inevitably the lost continent of Atlantis, populated at the time by the Asian colonial peoples. Atlantis and cosmic ice crystals led Himmler to embrace a bizarre cosmological theory of mediaeval aspect, called the *Welteislehre*, or world ice doctrine. The *Welteislehre* was conceived at the turn of the century by an Austrian engineer named Hanns

Hörbiger, a harmless eccentric and also the founder of a famous Viennese theatrical dynasty. (Two of his sons, Paul and Attila, were matinée idols in the years between the two world wars, and Paul Hörbiger's daughter was also a celebrated actress, but two other sons

HÖRBIGERS

GLACIAL-KOSMOGONIE

EINE NEUE ENTWICKELUNGSGESCHICHTE DES
WELTALLS UND DES SONNENSYSTEMS

AUF GRUND DER ERKENNTNIS DES
WIDERSTREITES EINES KOSMISCHEN NEPTUNISMUS
MIT EINEM EBENSO UNIVERSELLEN PLUTONISMUS

NACH DEN NEUESTEN
ERGEBNISSEN SÄMTLICHER EXAKTER FORSCHUNGSZWEIGE

BEARBEITET
MIT EIGENEN ERFAHRUNGEN GESTÜTZT UND
HERAUSGEGEBEN
VON

PH. FAUTH

MIT 212 FIGUREN

„WAS DU ERERBT VON DEINEN VÄTERN HAST,
„ERWIRB ES UM ES ZU BESITZEN." GOETHE.

HERMANN KAYSERS VERLAG, KAISERSLAUTERN
1913.

Title page of Hörbiger's massive treatise on the *Welteislehre*, edited by his John-the-Baptist, Philipp Fauth. The legend translates as 'Hörbiger's Glacial Cosmology—A new developmental history of the universe and the solar system—on the basis of the recognition of the opposition of a cosmic Neptunism to an equally universal Plutonism—following the most recent results of all exact branches of research—revised and supported by his own experiences and edited by Ph. Fauth'.

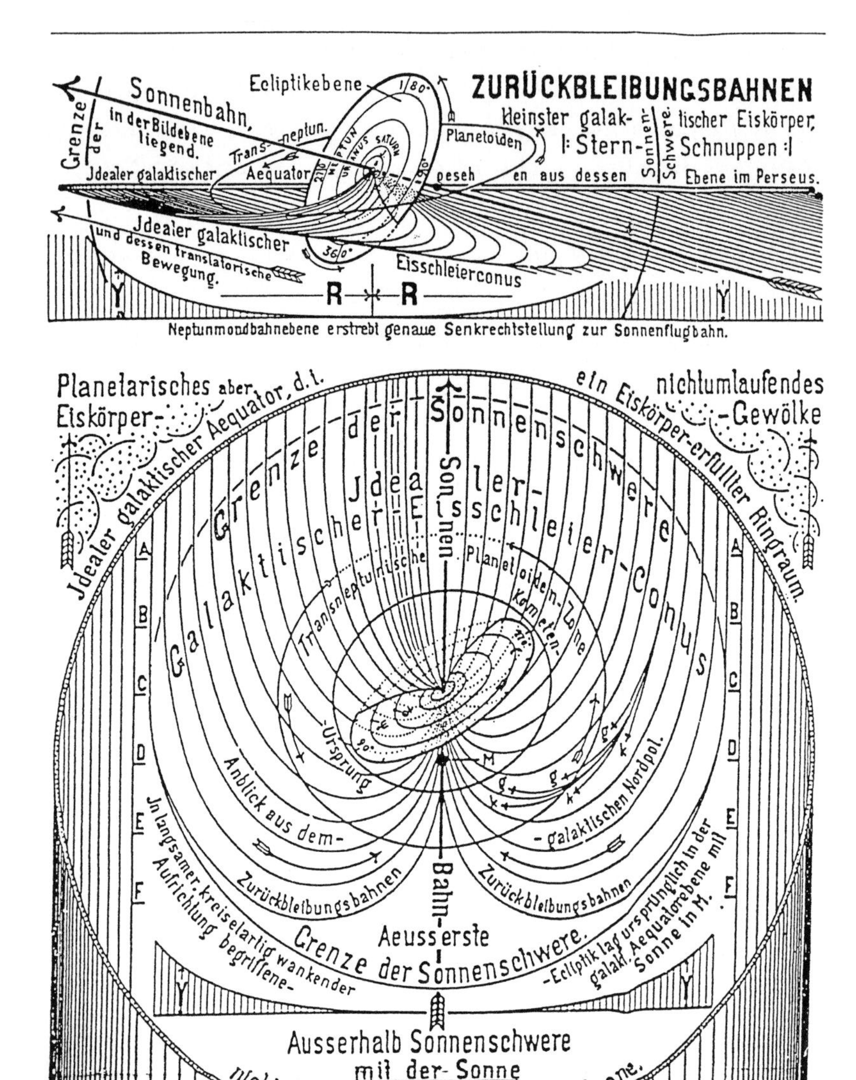

Explanatory diagram from Hörbiger's treatise.

devoted themselves to the promotion of their father's theory.) The new cosmology was revealed to Hanns Hörbiger in a dream, and elaborated through nights of spiritual turmoil. Hörbiger's most besotted follower was a German amateur astronomer, Philipp Fauth, whose book on the subject became the accepted source of wisdom.

In essence the idea was the following: the planets are covered by a thick layer of ice, and even the moon has a kilometre-deep encrustation. The milky way is nothing more than a mass of ice crystals. When such cosmic ice enters the gravitational field of the sun, it is drawn in and explosively vaporised. The sun-spots are funnels of exploding steam, great fountains of which—the solar prominences—are then ejected back into space. There the steam instantly freezes into crystals or larger lumps, known to us as meteors. The ice crystals rain down on the planets, adding to their icy crust. When they enter the earth's atmosphere the crystals melt and turn into rain, while the meteors explode and shower us with hailstones. The weather therefore comes from outer space.

But then Hörbiger has a more apocalyptic vision: the drag of the luminiferous aether, through which the heavenly bodies travel, is all the time slowing them down, and some will in time be captured as moons by the inner planets. The moons, of course, will spiral towards the planets and eventually crash into them, just as happened when, aeons ago, the earth's second moon fell into the ocean and became the continent of Atlantis, and, for good measure, engendered Noah's flood. The end will come when the earth in its turn is drawn into the sun and consumed in fire. Hörbiger noted the consonance of his theory with the Nordic myth of the earth's creation out of fire and ice. The appeal to a visionary of Himmler's credulous and impressionable temperament was all too obvious.

The *Welteislehre* found considerable favour among the ruling élite of the Third Reich. Hitler said that he was 'inclined to believe' that the theory was correct, and Göring, Speer and the head of the Hitler Youth, von Schirach, were all thought to be strongly drawn to it. The Reichsführer SS gave instructions that a study of long-range weather patterns, in relation to the sun-spot cycle and astronomical events, should be urgently undertaken by the Ahnenerbe to obtain proofs of the correctness of the theory. Accurate long-range weather forecasts, which the predictions of the *Welteislehre* were expected to encompass, would moreover have obvious practical value. As head of the weather section an Air Ministry meteorologist, Hans Robert Scultetus, was appointed, and appeared to need little convincing that the theory held all the answers. The latter-day champion of the world ice, Philipp Fauth, was also recruited, complete with his private observatory in Munich.

But it was not long before the consternation of German physicists—including those with Nazi convictions—broke surface at the spread of this balderdash. The attack was led by the veteran Nazi and Nobel Laureate Philipp Lenard, who took the occasion of an article (intended as the first of a series) on the *Welteislehre* in a Nazi paper, the *Illustrierter Beobachter*, to address a fierce reproof to the editor. Was a National Socialist paper to be allowed to force such asininity on the German people? The theory was pure fantasy, a perversion of our knowledge of the world. The editor was cowed and pulled the planned articles, and Scultetus and Himmler fumed. Both hated and feared the professoriat, whom they called 'the calcified bonzes', and they debated how best to combat them.

It was a losing battle, however. In a deposition to the Ministry of Education, the director of the Berlin University observatory heaped scorn on the preposterous theory, which he described as a mediaeval throwback. To call this kind of science typically German was too much. It could be better described as the product of Bolshevik or scientifically subhuman minds. It would be interesting, he continued, to discover who financed the undertaking. It was, he thought, a matter of urgency to dissociate the person of the Führer from this shameful affair. Himmler reacted with fury. He demanded that the Reich Education Minister reject all such arrogant effusions from professors. Intimidated, the Ministry divulged that the report had been commissioned by a member of Himmler's own staff, by name of Polte, and so delivered into Himmler's hands a victim when he most needed one. Polte was dismissed from the SS and officially disgraced. The affair in fact had wider ramifications, as will emerge.

The hatred that Himmler and his myrmidons bore towards the scientific establishment was tempered by fear of its power of argument, and research on the *Welteislehre* henceforth went underground, covertly pursued in corners of the Ahnenerbe under Himmler's constant prodding. He demanded, for instance, an investigation into the relation of 'atmospheric electricity' to disasters, such as the destruction of the dirigible, the *Hindenburg*, in 1937. The idea came to him that studies of mineral deposits would reveal traces of the workings of the *Welteislehre* and expeditions to Asia and South America were planned to search for such evidence. These were postponed when the war began. A proposed expedition to Iceland met the same fate: it was to have brought back evidence for the direct descent of the leaders of the Third Reich from the Vikings.

Folklore, tradition and the Vikings were always close to the Reichsführer's heart. Here is a letter written by Wolfram Sievers to a Fräulein Piffl. (It is reproduced by Sam Goudsmit, in his book, *Alsos.*[2])

> Dear Fräulein Piffl—There was a recent report in the press that there is an old woman in Ribe in Jutland, who still possesses knowledge of the knitting methods of the Vikings. The Reichsführer desires that we send someone to Jutland immediately to visit this old woman and learn these knitting methods. Heil Hitler!

The preoccupation with the weather and its control by the migration of the cosmic ice also continued. As late as 1944, when the Third Reich was already marked for destruction, Himmler wrote to Sievers a letter, also discovered by Sam Goudsmit's investigators. This is how it read:

> In future weather researches, which we expect to carry out after the war by systematic organisation of an immense number of observations, I request you to take note of the following: 'The roots, or onions, of the meadow saffron are located at depths that vary from year to year. The deeper they are, the more severe the winter will be; the nearer they are to the surface, the milder the winter'. This fact was called to my attention by the Führer.

Although its headquarters were located in Berlin, outposts of the Ahnenerbe were established in several centres, including Vienna after the *Anschluss*, and, when the war came, in the conquered territories of the east. Sections on geology, speleology, palaeontology, anthropology and geography, and on animals and plants were set up and the philology section was expanded. There was also a section on 'secret', presumably occult, sciences. As an outlet for the results of the Ahnenerbe's researches a journal was founded, its title *Germanien.*

With the coming of war the Reichsführer's brainwaves gave rise to new research projects with a more practical turn. He appointed an obscure zoologist, Ernst Schäfer, head of natural sciences and instructed him to set up programmes on the nutritional methods of the ancients, which might mitigate food shortages. Among the traditional methods that Himmler wished to see revived was the brewing of mead from honey. Some bottles of the life-giving liquor were produced, but it was judged that honey could be put to better nutritional uses. Farms were established in the concentration camps to cultivate medicinal herbs, and angora rabbits were bred for wool to make underwear for the soldiers on the Russian front. The rearing of horses for food was another *idée fixe* of Himmler's. He had read that the Mongol horsemen who overran Eastern Europe had benefited from a diet of dried horse meat and frozen horse's milk, which they carried in their saddle-bags. The cultivation of ancient native plant species, including trees, was encouraged. These were held to be superior to modern stock, being straight, tall and unyielding, not like for instance the beech, warped and therefore degenerate. A herbal garden of 200 acres was planted in the Dachau camp, tended by at least a thousand prisoners.

From his special train in Poland the Reichsführer sent a message to Professor Schäfer that he had spotted in a field a brown horse with a white mane: was this perhaps related to the red horse with white mane that featured in a Nordic legend? Schäfer did not know, but soon he was given another equine poser—to breed a superior steppe-horse, like the wild reddish-brown Przewalski's horse, which would serve soldiers and settlers throughout the severest winters as a draught-animal, a mount and a source of meat and milk. A stud-farm was organised in Norway and a herd of Eastern European horses was also transported to the Buchenwald concentration camp. This comedy was played out in 1945 and cost the lives of a number of SS men before the surrender of Germany brought it to an end.

It was inevitable that the pervasive German pseudo-science of *Rassenkunde* should find a place in the Ahnenerbe. Himmler had his own ideas, of course, and sought a collaboration with the State Office on Race. The primitive races of the south, he decided, had been annihilated by the Nordic warriors, but descendants survived as Hottentots and Jews, for it was a matter of observation that steatophygia, the overdeveloped buttocks seen in some African tribes and in their figurines, was also a characteristic of Jewish women. An anthropologist, Bruno Beger, was enlisted for these studies, and was eventually let loose in Auschwitz. Other races to be investigated included those of Siberia, in which Beger had a special interest, but, with the retreat of the German armies in Russia, the best that could be done was to make captured Yugoslav partisans available for study. Himmler also cherished a theory that the archetypal Nordic warrior was distinguished by a Grecian nose. He therefore formulated a project for breeding from individuals thus endowed, but, aware presumably that the consummation of this plan would take time, he demanded that SS men with Grecian noses should be sought out and formed into a special Waffen SS unit, where their comportment in battle could be observed.

These crackpot undertakings were for the most part relatively harmless, if at times detrimental to the German war effort, but the begetter of the extermination camps was not backward in finding uses for the 'human material' now so copiously available. By no means all of the experiments on the victims of the concentration camps fell under the aegis of the Ahnenerbe. As we have seen, the anthropologists, such as Fischer and von Verschuer, found willing helpers with access to the prisoners in Auschwitz and elsewhere, of whom Josef Mengele was the best known. Mengele was intensely ambitious for an academic career and had already acquired doctorates in both medicine and physical anthropology. He indulged his interest in anatomical measurement, as well as performing ludicrous and brutal experiments, such as injecting the eyeballs of brown-eyed prisoners with blue dye, to see whether the colour of their irises would change. Human eyes, especially from gypsies, were also despatched in large numbers to Verschuer and his colleagues at the Kaiser-Wilhelm Institute for Anthropology, Human Heredity and Eugenics in Berlin. This was only a part of the continuing traffic in body parts and blood samples between Mengele and his academic collaborators. After the war Verschuer, despite abundant evidence of his complicity, denied any knowledge of the provenance of these grisly materials and a committee of professors declared him innocent of all impropriety; they somewhat marred their case by describing Mengele as a caring camp sick-bay doctor, but Verschuer serenely resumed his academic career as Professor and Director of the Institute of Human Genetics at the University of Münster.

Himmler's main rival in the party, its acknowledged political theorist, Alfred Rosenberg, had founded his own Institute for Research into the Jewish Question, and he was by no means the only other Nazi leader with intellectual pretensions. Even the deranged Julius Streicher formulated the theory, which he expounded in his hate-filled newspaper, *Der Stürmer*, that cancer was a bacterial infection, and ordered a research project on the question. Himmler, as an adherent of what we would now call alternative medicine, had his own agenda. He did not on the whole care for real doctors and was himself attended by an unqualified 'naturopath'. At the Neuengamme concentration camp, inmates were infected with tuberculosis, so that in accordance with homeopathic principle (Chapter 7) they could be treated with low doses of tubercle bacilli. The results can be imagined, but the experiment was continued for years to provide its organiser with enough data for a thesis.

One of Himmler's preoccupations concerned the spread of disease by insects, and in particular he wanted to know why lice and other pests attack one man and not another. He surmised that those of

Nordic race were resistant and he wanted party leaders to demonstrate as much in insectariums that he proposed to build: this scheme found little favour among his friends. A unit was set up in Dachau, however, under a zoologist commissioned by Sievers, to breed fleas, lice, bedbugs, mosquitoes, horseflies and houseflies and release them on the experimental subjects. Many of the leadership also went in fear of 'earth rays', which the superstitious believed to be the cause of diseases. These could be detected, like underground water, by skilled dowsers, who were often engaged to seek out the lethal fluxes in public buildings. Hitler's deputy, Rudolf Hess (who lived to the age of ninety), would sleep with magnets hung above his bed to neutralise dangerous magnetic radiations that he had absorbed during the day.

Anatomy was a fruitful area, and Himmler's protégé Bruno Beger, who had been engaged to study racial anatomical characteristics, was seconded to assist Professor August Hirt of Strassburg University by procuring 'materials' from the camps. Hirt was an anatomist of some repute, who had been installed at the newly repossessed university in the city that the Germans once again claimed for their own. Hirt was a sinister creature; his very appearance was frightening, for much of his lower jaw had been shot away in the first World War. He saw the death camps, and indeed the war of extermination in the east, as an opportunity for acquiring valuable human anatomical specimens. His programme seems to have been an ambitious one, and he wanted especially to establish a collection of skulls in Strasbourg. He sent a demand for Jewish skulls in large numbers and, to procure the most transcendentally pathological skulls and brains, he requested that a doctor be attached to the German armies in Russia, who would look out for captured Jewish Bolshevik Commissars. These were to be carefully killed, so as to ensure that the heads were not damaged. The heads were then to be detached and preserved for anatomical measurements, both in the field and in Strasbourg. The army in Russia seems to have been too busy with other priorities to indulge Professor Hirt, and he had to make do with concentration camp specimens.

In 1942 Sievers wrote an order for 150 skeletons of Jewish prisoners in Auschwitz. Beger, with a team of anthropologist assistants, selected the victims and transported them to the concentration camp at Natzweiler, near Strasbourg, where they were gassed according to Hirt's instructions and anatomised. Representatives of other races were also studied, and a number of doctors and anthropologists were offered helpful courtesies by the SS. Remains of the *Untermenschen* were gratefully received by other institutions and some are undoubtedly still held in German anthropology departments. Quite recently a collection of skulls supplied from a concentration camp was discovered in the Natural History Museum in Vienna. The nineteenth-century pursuits of phrenology and 'craniometry'—the topography and dimensions of the cranium (p. 156)—also enjoyed a resurgence in these circles, having been used in earlier years to seek out racial stereotypes.

Much of the most brutal experimentation on human subjects was carried out under the direction of an arm of the Ahnenerbe, established on Himmler's orders well into the war. This was the so-called Institute for Applied Research in the Natural Sciences *(Naturwissenschaftliche Zweckforschung)*. One of the most repellent of the psychopaths who directed these fatuous programmes was an SS doctor by the name of Siegmund Rascher. It was he who conceived the infamous 'altitude' experiments, in which prisoners in Dachau were subjected to low pressures. This project, which was not endorsed by the doctors of the Luftwaffe, claimed the lives of 80 per cent of the experimental subjects. Rascher followed this inquiry by an equally sadistic and futile investigation into the effects of cooling. This time the director of the Luftwaffe medical services, Colonel-General Professor Dr (for punctilio about titles went with the love of insignia and polished leather in the Third Reich) Hippke, gave his approval and commissioned a physiology professor from Kiel University to assist. The ostensible purpose of the project was to discover how to revive moribund sailors or airmen who had been immersed for long periods in icy waters. Contact with a warm human body (female) might be the answer. Himmler was enthusiastic about this research, giving as his reason that 'everybody knows that animal heat works differently from artificial heat', and he came himself to Dachau to observe the experiments in progress. The hypothermic and comatose subjects were put to bed with one naked woman, or sandwiched between two women. (Prostitutes were procured for the purpose.) If the victim revived, sexual activity might even result, and this was obviously of keen interest to the voyeuristic Himmler and Rascher. Control experiments were done with no women in the bed and also with hot baths to effect warming. Rascher submitted the results of his researches in a habilitation thesis (a dissertation, still required by German universities, as proof of the supplicant's mastery of his subject and fitness to teach it at university level), but the examiners would have none of it. Yet not one of the university professors in the audience voiced any reservations when Rascher gave an account of his experiments at a medical conference. It is satisfactory to record that Rascher met a nasty end under the system that he had so odiously exploited: his wife had borne him no children, but because a large brood of male offspring was a great career advantage in the SS, he had abducted a series of babies, each time sending Himmler news of another happy event. He was found out, tried, charged with the further crime of taking bribes from prisoners, incarcerated in Dachau and in due course shot.

Another of the pseudo-intellectual scientific servants of the SS was Joachim Mrugowsky, a doctor and SS member since 1931. His mystical and vitalist beliefs, articulated in papers and a book, had incurred the opposition of the racial geneticists, such as the influential Fritz Lenz, who blocked his path to the academic heights of a professorial chair in Berlin. Such disapproval, however, was no bar to his rise in the ranks of the SS and in 1942 he was appointed Commissar for Epidemics in the Ostland—the conquered territories in the east. The SS had already organised research centres for the study of typhus and enteric fever in concentration camps, especially in Buchenwald. Now they enlisted the help of the giant pharmaceutical concern, Behringwerke, which had established an institute in the Polish (now

Ukrainian) city of Lwów (to the Germans, Lemberg). Mrugowsky supervised trials with Behring sera, but he also believed that the inferior races were especially susceptible to epidemic disease and were responsible for infecting their betters. Eugenic measures must therefore be undertaken to stem the spread of epidemics. These activities and many experiments on deliberately infected civilians cost Mrugowsky his neck when he was tried after the war.

Many German biologists may have viewed Rascher and Mrugowsky and their like with distaste, but the intellectual quality of their research was little inferior to much that was going on in the most venerable centres of learning. Wolfgang Abel, for illustration, Professor of Anthropology at Berlin University, ran a programme on sex determination of babies: 'We know from old information and population-political statistics that in time of war the balance of births shifts to male'. Abel was especially interested in the anthropology of the Russian race, and the menace that the Russians presented to the future of the *Volk*. The problem had to be faced: the Russian people must be exterminated, or, at the very least, the Nordic territories must be Germanicised, and the indigenous population driven out. Then there was Siegfried Passarge, a geographer at Hamburg University, who became the representative of his discipline in the professional association of university teachers and scientists. He wrote eloquently about the superiority of the *Langschädeln*—the long-skulls, that is to say the Aryans—whom he described as the movers, the dynamic people. These he contrasted, much to the detriment of the latter, with the *Rundschädeln*, or round-skulls, who were sessile and indolent, 'in short, democrats'. Such were the torrents of bilge flowing from the academies of a country that so prided itself on the effulgence and rigour of its scholars.

Die deutsche Physik, its friends and enemies

The intellectual leaders of the Third Reich had grandiose ambitions; it was their aim to create a new integrated National Socialist culture:

there must be a new literature, theatre, cinema, painting and sculpture, architecture and design. It was characterised, when it came about, by a uniform gigantism, as in the architecture of Albert Speer and muscle-bound nudes of Arno Breker. In 1936 Munich was treated to the famous exhibition of *entartete Kunst*, degenerate art, in which were displayed works by most of the significant artists then active in the West, which visitors were invited to deride. Opposite was a gallery exhibiting the art of the Third Reich, which they were expected to admire. It is scarcely surprising therefore that similar attempts were made by the scientific adherents of the regime to develop a National Socialist ideology of science. For biologists it was easy: the racial axioms had already been enunciated for them. In physics and mathematics the solution was less obvious. It emerged eventually from personal jealousies and from the strong representation of Jews in physics, more especially the new physics of relativity and quantum mechanics, which some classical physicists found deeply disturbing and even threatening. (The difficulties that many honest scientists experienced with these developments, which appeared to them to undermine the certainties on which their intellectual creed was based, are penetratingly but sympathetically described by Russell McCormmach in his historically based novel, *Night Thoughts of a Classical Physicist*, set around the time of the first World War.)

The founders and indeed the only influential adherents of the movement known as *Die deutsche Physik* were two Nobel Laureates, Philipp Lenard and Johannes Stark. Both were, or became, rabid anti-Semites, as well as being temperamentally hostile to the new physics, which they primarily identified with the persons of Einstein and Heisenberg. Lenard was the older and more interesting of this turbulent pair, though undoubtedly in the last two decades of his life more than a little mad. He was born in 1862 into the family of a prosperous wine merchant in what is now Bratislava, the capital of Slovakia, and was then the Hungarian city of Poszony (but known to the Germans as Pressburg). Lenard became in his youth an ardent Magyar nationalist. His preferred language was Hungarian and he

Philipp Lenard, originator of *deutsche Physik*. AKG London.

objected violently to the use of German place names in the predominantly Hungarian province in which he lived. He styled himself Fülöp Lenard, or according to later rumour, Lenardi. There is no evidence that he was at that stage of his life anti-Semitic, and he revered the half-Jewish Heinrich Hertz, as whose assistant he served at the University of Bonn.

Johannes Stark, champion of *deutsche Physik* with Philipp Lenard. Science Photo Library.

Lenard was an experimental physicist of great ingenuity and imagination. He began to work on cathode rays under Hertz and made many fundamental measurements on their properties. An important innovation was the 'Lenard window', through which the beam in the cathode-ray tube could be directed onto targets on the outside. In a remarkable case of simultaneous discovery, Lenard became one of four physicists who independently determined that the rays were negatively charged, though he always asserted his own priority. He showed that the cathode rays—free electrons—can pass through atoms, and he inferred that much of the atom was empty space, in which positive and negative charges were separately clustered. But it was left to Rutherford to put forward the planetary structure of the atoms, comprising negatively charged electrons revolving in orbits around an electropositive nucleus. It was also an enduring grief to Lenard that he missed discovering X-rays: he could never in later years bring himself to pronounce the name of Röntgen. In 1905 Lenard was awarded the Nobel Prize and in 1907 he was called to the

Chair of Experimental Physics at Heidelberg, where he remained for the rest of his career.

A prickly and mistrustful man, who carried grudges to the grave, Lenard's life was also embittered for many years by a distressing disability, a painful muscular disorder which made it impossible for him to raise his head from a hunched position. It was only after a long succession of failed treatments that an operation eventually succeeded. Despite his sour nature, Lenard was by all accounts a remarkable teacher and lecturer, who would preside like a conjurer over his demonstrations: when the experiment produced the desired flash or bang, Lenard would leap back theatrically with a cry of 'Ah!' But he never developed a thriving research school, and in his obituary in *Nature* the British physicist E.N. daC. Andrade thought that he often appeared a pathetic rather than a menacing figure.

What ignited this strange man's violent animosity against Einstein, whom he had earlier held in high regard, is a mystery. It may have been an imagined slight, or the gradual realisation that the kind of classical physics he represented and loved was being pushed to the margins by the relentless advance of theory, with Einstein in the van; it may even have been merely Einstein's elimination of the aether from physics, and with it a seeming repudiation of common sense; or it may have been Einstein's loudly proclaimed abhorrence of the Great War, which in Lenard had evoked an ebullition of patriotic fervour. Even there, though, *amour propre* appeared to have played a part: James Franck, a noted experimental physicist (and later a winner of the Nobel Prize and an early emigré from the Third Reich), received, while soldiering in France during the war, a letter full of martial encouragement from Lenard. According to Franck, Lenard had written 'that we should especially beat the Englishmen, because the Englishmen had never quoted him decently'. This grudge probably stemmed mainly from an altercation in 1903 with the Cavendish Professor of Physics at Cambridge, Sir J.J. Thomson, who had sought in a paper, so Lenard believed, to deprive him of credit for his work on the photoelectric effect. Lenard had in fact published a pamphlet at the outbreak of war accusing the English of inborn dishonesty, in science as in political matters.

Lenard was evidently disgusted at the public exposure that Einstein, widely acclaimed in the period after the war as the world's greatest physicist, was getting. Einstein was himself uncomfortable. In a letter to his friend, Max Born, he wrote that his Midas touch was turning everything, not to gold, but to newsprint. Lenard now associated himself with the anti-relativity, pro-aether faction of vocal but largely insignificant physicists, and became, by reason of his scientific stature, their leader. Einstein, provoked, made fun of them, and of Lenard in particular, in a popular article. The confrontation came at the 1920 meeting of the German Physical Society in Bad-Nauheim, where Lenard attacked Einstein in intemperate, and for the first time anti-Semitic, terms; he attacked the theory of relativity as counter-intuitive—the word *anschaulich*, meaning evident or transparent, was heard and became the rallying cry of the 'Aryan physicists'; he

Albert Einstein

attacked Einstein for debasing a scientific debate by resorting to vilification of his opponents, 'a well-known Jewish trait'; and he attacked Einstein's friends for supporting Einstein's science for reasons of racial solidarity.

Lenard, whose excesses dismayed the moderate majority of the scientific community, sank ever deeper into the mire. In 1922 Walter Rathenau, the Jewish Foreign Minister of the Weimar Republic, was assassinated by fascist thugs. Einstein received death threats, which he took seriously enough to stop appearing at public events. Lenard's already unstable state of mind was aggravated by the death of his son and the loss of his savings, with the collapse of the national currency—a catastrophe imputed by the political right to the machinations of the Jewish financiers, who, they believed, controlled the economy of the Weimar Republic. A national day of mourning for Rathenau was ordained by the government, but Lenard let it be known that he applauded the murder (and that in the previous year of the left-wing politician, Matthias Erzberger) and he refused to close his Physics Department at Heidelberg or fly the flag at half-mast. A large group of left-wing student demonstrators marched on Lenard's laboratory and confronted him. Barricaded into the building, exposed to the insults of the students and afterwards censured by the university administration, Lenard's world collapsed. He was never able to forget the affront to his dignity and he lapsed into a state of paranoia. Jewish and English conspiracies were all around him. He abandoned the Church, because it exalted the apostles, who were Jews. If Max Planck, the presiding luminary of physics in Germany, supported Einstein (and he was indeed a friend of Einstein's, with whom he enjoyed performing chamber music), Lenard believed it could only be because he came from a line of churchmen, with a misguided respect for the Old Testament and therefore for the Jews. In 1925 the *Zeitschrift für Physik* published a paper from a British laboratory in English, and Lenard promptly resigned from the German Physical Society, putting up a notice at the door of his department, which announced that entrance was denied to members of the Society.

Lenard began to make contact with racist biologists and one of them, H.F.K. Günther (whom we have already encountered), put into his mind the idea of a study of the great Aryan physicists. This duly appeared under the title of *Grosse Naturforscher*, and even the British physicist E.N. daC. Andrade (a Sephardic Jew by origin) called it a masterly survey in his review in *Nature*. Lenard then embarked on his four-volume treatise, *Deutsche Physik*, in which he laid down the principles of true physical science: it must be *anschaulich*, intuitive and *blutmässig*, or racially fitting. In 1927, at sixty-five, Lenard announced his retirement, but withdrew it when he learned that the short-list of proposed successors consisted of Jews and anglophiles. Two years later he finally took his leave, withdrew to his home in the country and took up his pen to make more trouble and to eulogise Hitler and the rising Nazi party.

At about this time began Lenard's association with that other champion of *deutsche Physik*, Johannes Stark. Twelve years younger than Lenard, Stark had been awarded the Nobel Prize for physics in 1919. His contributions had been the discovery of the 'Stark effect', and before that of the Doppler effect in canal rays.[3] Because of his personality, Stark had a hard time gaining a foothold in the German academic system, and it was 1917 before he succeeded to a professorial chair at the University of Greifswald, moving three years later to Würzburg. His colleagues there looked askance at his growing concern with technological problems. He was in fact becoming increasingly absorbed in the technicalities of ceramic manufacture, for with his Nobel Prize money he had bought a partnership in a porcelain factory. Such sordidly commercial research topics were deemed inappropriate for a university department; nor was he considered to be acting properly in using the money for investment, rather than for the furtherance of fundamental research, as Nobel's will specified. Stark had also developed a fanciful and generally derided model of the atom, in which the electrons, themselves rings, formed a doughnut-shaped ring around the nucleus. The opinion of Einstein's friend and associate, Max Born, was that 'Stark was a genius at fiddling with apparatus, but he had no understanding of physics'. The low regard in which he was generally held by other physicists undoubtedly fed Stark's resentments.

Stark was the archetype of the disagreeable man; he was vain, conceited, arrogant and quarrelsome. Like Lenard's, Stark's anti-Semitism developed late. He had admired Einstein, whom in the early stages of his career he had often consulted. Indeed, while an extraordinary professor (a grade below that of a full professor, or *Ordinarius*) in Aachen, he tried to persuade Einstein to join him there, and he had commissioned a review from Einstein on relativity for a journal that he edited. However, he soon fell out with Einstein, as was his wont, over a trivial issue of priority. The final breach came with the outbreak of the First World War, when, like Lenard, Stark fell into the grip of an ungovernable chauvinistic passion, and made no secret of his detestation for Einstein's pacifist posture (and refusal to sign the German intellectuals' appeal to the *Kulturwelt* (Chapter 8)).

The anti-Einstein movement, which the confrontation at Bad-Nauheim brought into the open, was to a remarkable degree associated with anti-Weimar and *Völkisch* politics. Its partisans were a minority, to be sure, but they did include, besides Lenard and Stark, a considerable number of scientifically reputable if not especially distinguished figures. Stark broadened the assault by delivering a series of open lectures, held in the Berlin Philharmonic Hall and rich in anti-Semitic rhetoric, directed mainly against Einstein. In 1920 and 1921 Stark engaged in a vigorous intrigue to gain control over German physics by starting a professional organisation of his own within the German Physical Society, and by seeking election to the chairmanship of one of the two major research funding bodies. He was opposed by all the leading figures in German physics, who put up a compromise candidate, the entrenched conservative, Willy Wien, for the presidency of the Physical Society to still the clamour from the right and deflect support from Stark's scheme. The stratagem worked and Stark's bid for power was thwarted. He retired in a huff to Würzburg but there, too, trouble awaited him, as the faculty registered its disapproval of his preoccupation with porcelain and of the ill-use to which he had put his Nobel Prize by rejecting the habilitation thesis of one of his assistants on the subject of optical properties of ceramics. (The candidate, mocked as a Doctor of Porcelain, became an active Nazi and was able to take his revenge when he attained academic power a decade later.) Stark, seeing conspiracy all around him, thereupon resigned his position and withdrew to his private laboratory on his country estate.

From his Bavarian fastness there flowed a series of articles and pamphlets, with titles such as 'The current crisis in German Physics', in which he denounced the physics establishment, the Ministry of Education and the politicians, and raged against the domination of theoretical over experimental physics and against relativity and quantum mechanics. It was not long before Stark began again to search for a university position, but, his Nobel Prize notwithstanding, he was repeatedly rebuffed. In 1930 he joined the Nazi Party (having already offered Hitler his personal friendship after the beer-hall *putsch* of 1923) and wrote a series of political pamphlets in its name. And then in 1933, with Hitler's accession to power, Stark's chance finally came. At last, he wrote to Lenard, the two of them would be able to implement their ideals of how science and research should be conducted. He sent a telegram to all Nobel Laureates in the country, inviting them to sign a manifesto, stating that 'In Adolf Hitler we, the German scientific researchers, recognise and admire the saviour and leader of the German *Volk*. Under his protection and with his inspiration, our scientific work will serve the German *Volk* and promote German prestige in the world.'

Stark made his revived ambitions known at the autumn meeting of the German Physical Society in Würzburg: he wanted to become director of the Physikalisch-Technische Reichsanstalt, the PTR. This was a state laboratory for applied physics and more or less equivalent to the National Physical Laboratory in Britain or the National Bureau of Standards in the United States. Secondly he coveted the presidency of what was now the principal body controlling the funding of research, the Notgemeinschaft (emergency association), which shortly afterwards became (and remains to this day) the Deutsche Forschungsgemeinschaft (the German research association, the DFG). This would give him control over the allocation of the bulk of research funds available to university departments. His most courageous opponent was one of Germany's leading physicists, Max von Laue, a man who never compromised with the regime and was the only member of the physics community to retain Einstein's esteem to the end of the Second World War. Rounding on Stark, von Laue accused him of wanting to become 'the dictator of physics' and likened his assaults (endorsed by the Nazi party) on the Theory of Relativity to the proceedings of the Inquisition against Galileo.

Nevertheless, that same year Stark was rewarded for his loyalty to the party with both the offices that he craved, the current incumbents having been unceremoniously turned out. He at once began to develop expansive plans for the reform of the PTR and of German physics generally. He demanded that a central agency should be set up to control all scientific journals, so that the material to be published could be censored. It seemed that Stark might indeed become dictator of physics. In 1934 von Laue publicly and eloquently opposed Stark's candidacy for election to the Prussian Academy of Sciences (the premier professional body for scientists in the country), even though Max Planck and others had allowed Stark's name to go forward, for the sake of peace and compromise. (The Prussian Academy, although it had complied meekly with the Ministry's instruction to expel all Jews, did continue to exercise some independence in the election of new members; this was also true of the Bavarian Academy of Sciences, which had the smallest proportion of party members—26 percent—but never of the Vienna Academy or the Leopoldina in Halle, which from the outset did as they were told.) Stark took his revenge the very next day by sacking Laue from his position as theoretical adviser to the PTR.

In the event, Stark's plans were too outlandish and his personality too abrasive for him to make much headway. His proposal for a new

Reich Academy of Sciences, with Hitler as president, incurred the displeasure of the Minister of Education, Bernhard Rust (whom he had already annoyed in other ways), and several powerful officials besides, notably the ambitious Rudolf Mentzel of the Ministry's Research Division. They did not wish to have their own power eroded by this noisy and truculent bore. Stark thereupon made a permanent enemy of Rust by accusing this impeccable anti-Semite of promoting the Jewish influence in science. In his capacity as head of the funding body, the DFG, Stark received, and predictably rejected, overtures from the Ahnenerbe for support of its research projects, and he may thereby also have alienated Himmler. He had in any case already aligned himself with Rosenberg, Himmler's great rival.

The expulsion of the Jews from academic and civil service positions had begun in 1933, with the initial exception of those who had fought in the Great War. The exodus that ensued of so many of the leading physicists from Germany, led by Einstein, the most internationally celebrated figure of them all, dismayed those who remained, such as Max Planck and Max von Laue. Planck, from whose work had emerged the concept of a quantum of radiation (the first large gash inflicted on the seemingly smooth and flawless surface of classical physics), was by then an old man. He had suffered much, for one of his sons had fallen at Verdun in 1916 and two daughters had died in childbirth. (His younger son was to be accused of complicity in the 20 July plot on Hitler's life and was executed shortly before the end of the war.) Planck nevertheless persevered in what he saw as his task, that of preserving what could be saved of German physics. As President of the Kaiser-Wilhelm Gesellschaft (renamed after his death the Max-Planck Gesellschaft), the scientific body that encompassed a nationwide chain of research institutes in all branches of natural science, Planck enjoyed considerable power and prestige. So in 1933 he had, as he recalled twelve years later, felt it his duty to confront Hitler with the devastation that the removal of the Jews was visiting on physics. (No less than a quarter of all physicists in academic institutions were dismissed or felt compelled to resign: the list is a veritable Who's Who of twentieth-century physics.) Planck had written to his friend, the great Dutch physicist, H.A. Lorentz, of the dismay and shame that he felt at the hounding out of Einstein, 'a man for whom the whole world envies us', by what he called a low intrigue. Planck left his own account of the visit to the Führer, which began amicably enough. Hitler declared that he had nothing against the Jews as such, but they had to go because they were all communists. As Planck began to remonstrate, the Führer proceeded to ramble: 'They say I suffer from weak nerves. It is a slander! I have nerves of steel!' He smote his knee with his fist and spoke faster and faster, rocking from side to side as his disseminated rage mounted. Planck realised that further intercession would be profitless and withdrew. It makes a good story, but there is some doubt about its veracity. Planck, who died a few months later, was eighty-nine when he set down his reminiscence, and he had only just recovered from a serious illness. The recollections of the most harrowing period of his life may have become hazy in Planck's tortured mind; at all events, friends in whom he confided soon after the event remembered his story quite differently. Planck probably did not broach the delicate subject of the Jews and Hitler in any case barely listened to what he had to say before impatiently turning his back. Planck seems in fact to have rejected calls from Otto Hahn and others for a public protest against the persecution of their colleagues, much as he lamented the consequences in private. He saw it as his duty instead to limit the damage that was being done to physics and therefore, as President of the Physical Society, to maintain good relations with the régime. There is a poignant description of Planck mounting the swastika-bedecked rostrum at a meeting of the society, twice half-raising his right arm and then finally giving the Hitler salute. Einstein in later years could not forgive Planck's decision to make no further public protest against the expulsion of the Jews and try instead to limit the evil from within the Physical Society. Asked in 1945 whether he had any messages for his friends in Germany, Einstein sent his regards only to von Laue.

Stark and Lenard, of course, were delighted at the cleansing of the academic stables. Stark deployed his talents for sophistry in defending the regime's policy in Western journals, especially *Nature*. He was answered with contempt by the physiologist (and Nobel Laureate) A.V. Hill and the mathematician G.H. Hardy. Lenard did not approve of Stark's action and scolded him for publishing in 'the Jewish journal, *Nature*'.

A small act of defiance was played out in 1935 after the death in exile of Fritz Haber: Planck disobeyed the order of the minister, Rust, to cancel a memorial ceremony for this great physical chemist (branded a charlatan by Stark).[4] Civil servants—a category that included members of all university faculties—were forbidden to attend the ceremony and newspaper reports of the event were suppressed. However, staff of the Kaiser-Wilhelm society came, as did industrialists and foreign diplomats, as well as the wives of the excluded professors. Solemn speeches were delivered and Planck paid tribute to 'this great scholar, upright man, and *fighter for Germany*'. There were few such episodes.

Fritz Haber. Science Photo Library.

Stark's reign as head of the DFG ended in 1936 amid the customary acrimony. His dismissal was precipitated in part by a controversy over an ill-starred and expensive project for extracting gold from the south-German marshes, but more especially by a serious political altercation with the party Gauleiter of Stark's home province in Bavaria. Stark accused this official of peculation, was falsely accused in his turn and lost the court case that ensued. The affair dragged on intermittently well into the war years and Stark was eventually vindicated, but it was a Pyrrhic victory, for his action was held to have been damaging to the party. Stark's successor at the head of the DFG was Rudolf Mentzel, whom he held in well-merited contempt. There still remained to be enacted the contretemps between Lenard and Stark on the one hand and Heisenberg and Himmler on the other.

This ludicrous affair arose out of the insistence of the two zealots that there existed a pure transcendent Aryan physics, distinct from the trash produced by lesser breeds, such as the Jews and the French. The criteria by which it could be recognised bore a remarkable similarity to those laid down in Russia for Marxist physics (Chapter 9). The *deutsche Physik* must be based on experimental observation; it must be clear and *anschaulich* (as defined above), in tune with, and humble before, nature, the innermost mysteries of which would remain forever beyond human reach, and above all pragmatic. It was tempered in the culture of the race. The course of physics had been polluted by what Lenard in his treatise on *deutsche Physik* called *Stoffwahn*, or materialistic mania. Newtonian physics, which formed the bedrock of the discipline, had been debauched by the degenerate adherents of the French Enlightenment, and so on. Newton's maxim, *hypotheses non fingo*, was shamefully disregarded. The Jewish spirit was inimical to all these principles, for it elevated abstract, mathematical thought above observation; it spurned physical evidence, was dogmatic and arrogant and mirrored the racial characteristics manifested in business—the elevation of profit-and-loss calculations above ethics. No good had ever come from Jewish physicists, or, if it had, it was only because of admixtures of Aryan blood. Thus

Heinrich Hertz, Lenard's patron, had been a great experimental physicist—he had discovered radio waves—thanks to his Aryan mother, but when he turned to theory later in his brief life, it was the blood of his Jewish father asserting itself. Most importantly, the new physics of relativity and quantum mechanics must be expunged if German science was to remain pure.

What troubled Lenard and Stark was that these Jewish concepts had also been embraced by certain Aryans. These they dubbed 'white Jews' and the chief reprobate was Werner Heisenberg. That Heisenberg was one of the great physicists of all time cannot be doubted. A prodigy, he was awarded his doctoral degree at the age of twenty-two; at twenty-five he received the call to the chair of theoretical physics at the University of Leipzig, and in 1933, when he was thirty-two, he was awarded the Nobel Prize. It was Heisenberg who fashioned quantum mechanics into the most powerful theoretical tool in twentieth-century physics. He had worked with Niels Bohr in Copenhagen and had struck up a close personal and scientific rapport with the older man. Heisenberg's first great achievement was to deduce from the line spectrum of molecular hydrogen (H_2) that it existed in two forms (called ortho- and para-hydrogen), in one of which the nuclei spun in the same direction about their axes and in the other in opposite directions. Heisenberg is best remembered now for enunciating the celebrated Uncertainty Principle,[5] which put an end to the fully deterministic view of the universe associated with the name of the Marquis de Laplace. Laplace held that if one knew the positions, speeds and directions of motion of all the particles that make up the universe, its past and future could in principle be exactly calculated. (When Napoleon asked Laplace how God entered into this picture, he received the answer 'Sire, I have no need of that hypothesis'.)

In 1935 one of the mandarins of theoretical physics, Arnold Sommerfeld, under whom Heisenberg had studied at Munich University for his doctorate, reached retirement age, and he and the Munich faculty indicated that Heisenberg was their choice for

Werner Heisenberg. DHA/NMPFT/Science & Society Picture Library.

successor. Stark and Lenard were incensed. They had considerable support in the party, and especially from Alfred Rosenberg. The political complexion of the junior faculty and student bodies in the universities had undergone an enormous change. In Heidelberg, ten years after Lenard's traumatic experience at the hands of the socialist students, there was a torchlight procession (a high tribute) in his honour. The Philipp–Lenard physics institute was inaugurated in a ceremony at which Stark and Lenard made inflammatory speeches; the proceedings concluded with *Sieg Heils* and the Horst Wessel song.

But the monolith of the Nazi Party was deeply fissured by personal animosities and rivalries. Stark had in any case annoyed too many important functionaries, especially Rust and his officials at the Ministry of Education, who now aligned themselves with the majority of the country's leading physicists in opposing their bugbear. Stark's supporters were mainly minor academics, many of them former students of his own or of Lenard's, several with high positions in the SS. A group of these was appointed by Rosenberg as science consultants to the editor of the official Nazi party newspaper, the *Völkischer Beobachter*, and together they concocted an article about Aryan and Jewish physics, in which Heisenberg came in for much obloquy. Heisenberg was allowed a rejoinder, but this was followed in turn by a vitriolic concluding statement from Stark.

Stark now went for broke. Despite the known animus between Rosenberg and Himmler, and the rebuff that he had offered the Ahnenerbe while still head of the DFG, not to mention the contempt with which he and Lenard had treated the *Welteislehre*, Stark boldly ascended the favoured pulpit of the SS itself—the journal, *Das schwarze Korps*. In its columns in July of 1937 Stark launched his famous attack on the Jews in physics and especially the 'white Jews'—those Aryans in the midst of the German physics community who harboured the pernicious 'Jewish spirit'. It was not, Stark declared, the Jews who had to be fought, but the spirit that they diffused.

'And if the bearer of this spirit is not a Jew, but a German, he is all the more to be combated than the racial Jew, who cannot conceal the fount of his spirit. For such carriers of infection the voice of the people has coined the description of "White Jew", which is particularly apt, because it broadens the conception of the Jew beyond the merely racial. The Jewish spirit is most clearly discernible in physics, where it has brought forth its most "signficant" representative in Einstein.'

This philippic continued for a full page (though it was thought that other hands may also have contributed), and the most dangerous assault was on Heisenberg. 'In 1933', Stark wrote, 'Heisenberg received the Nobel Prize, together with the Einstein disciples, Schrödinger and Dirac [respectively Austrian and British, and neither a Jew]—a demonstration by the Jewish-influenced Nobel committee against National-Socialist Germany, comparable to the award to Ossietzky.' (Carl von Ossietzky was an outspoken pacifist journalist, who was awarded the Nobel Peace Prize while imprisoned in a concentration camp, where he later died.) Stark henceforth referred to Heisenberg as 'the Ossietzky of physics' and suggested that it would be best if he were to disappear 'like the Jews'.

Heisenberg had failed to get the chair at Munich, despite the strong representations of the faculty, mainly because of opposition from the student body and the professional association of university teachers, and he was now in deep trouble, having also been accused of earlier dismissing an Aryan assistant in favour of a Jew. The faculty gave in and prepared a list of other worthy candidates for the theoretical physics chair, but all were rejected in favour of a nonentity, who had distinguished himself by writing the previous year a tirade against Jewish science. The Dean of the university exulted that the degenerate theoretical physics of Einstein and Sommerfeld (who was forbidden henceforth to lecture) had been blotted from the physics course (which in consequence became valueless).

To save himself from being driven out of science or worse, Heisenberg appealed to Heinrich Himmler for protection. The Reichsführer SS was a family friend: his father had been at school with Heisenberg's grandfather, both had become schoolteachers in the same small town, and their wives too were close. The approach was made in fact through Heisenberg's mother, and was met by a long silence. Not until a year later did Himmler write to Heisenberg, in somewhat patronising terms, stating that his case had been examined in exacting detail and exonerating him from any wrongdoing. Himmler added that he did not approve of the attack in *Das schwarze Korps* and he would forbid any such vilification in future. But he demanded from Heisenberg a promise of good behaviour. At the same time he wrote to his acolyte, the head of the SS Security Service, Reinhard Heydrich, that Heisenberg was to be protected in the hope that his talents would one day be put to use by the Ahnenerbe, and might in particular be enlisted to help with the programme of research on the preposterous *Welteislehre.* There is no record of any attempt to engage Heisenberg in such a project, but in 1942 he was appointed to a chair in Berlin and the next year to the presidency of the Kaiser-Wilhelm institute there, and of course he was chosen to head the half-hearted German atom-bomb project.

The *deutsche Physik* movement did not last much longer. Lenard was old and his mind was clouding, while his friend, Stark, had exhausted his credit. He had forfeited the support of his most powerful patron, Rosenberg, by consorting with Himmler and the SS, the detested Mentzel held the purse-strings of the DFG (with dismal results for science) and there was certainly no residual good-will in the Ministry. Stark continued to rage against the objects of his hatred. In a press conference he spoke as follows: 'Einstein has vanished from Germany. But unfortunately his German friends and champions are permitted to continue acting in his spirit. His main champion, Planck, still presides over the Kaiser-Wilhem Gesellschaft and his interpreter and friend, Herr von Laue, can still exert power in the Berlin Academy of Sciences [the new name of what had been the Prussian Academy]...'. But even his former supporters were wearying of this tiresome man. In 1939 Stark was finally dispossessed of his presidency of the PTR. He had become disenchanted with the corrupt party machine and with Hitler, who, he realised, had little regard for scientists or intellectuals of any kind; he wrote bitterly to Lenard that Hitler did not care about science. It was certainly true that such views as Hitler did have on the subject rested on a superstitious belief that nature knew best. He had apparently questioned the value of rockets on the grounds that they had no parallel in nature, and as for the Zeppelins, they had been a misbegotten idea, since after all birds had no lighter-than-air flight bladders.

No longer 'Johannes der Starke' or 'Giovanni Fortissimo' (as two of his adversaries, Gustav Hertz and Arnold Sommerfeld, had called him), Stark withdrew to sulk in his tent in Bavaria, to emerge in 1945 when the war ended, as assertive and self-righteous as ever. Until 1943 he had retained his party membership, which he had thought about renouncing in his dudgeon after the unfortunate court case, but his status as an *alter Kämpfer* (that is one who had joined the party before 1933) gave him valuable protection, and may have saved the life of his son, arrested by the Gestapo in the east on a charge of behaving humanely towards Polish labourers.

Stark's house had been occupied by the SS and, after they left, by the American army. He was brought before a denazification court to answer for his political activities during the Third Reich, and at the age of seventy-three was sentenced to four years' hard labour. The judgement was reversed on appeal, on the grounds that his actions had been professional, not political, in intent. His case was assisted by depositions from several associates, some even Jewish, whom he had defended. He indignantly rejected the offer of a mitigating statement by von Laue, but Heisenberg said that Stark's attack on him as a 'White Jew' was not personal. Einstein, asked for his opinion, wrote to the tribunal that in such questions of psychology an objective evaluation was impossible. Stark, he thought, was highly egocentric and had a powerful craving for recognition. His opposition to relativity theory and his political posture were a result of his paranoid streak. His behaviour, including his anti-Semitism had, in Einstein's view, been essentially opportunistic. But he, Einstein, had not suffered through Stark's activities. The court judged Stark to have been a minor miscreant in such evil times and fined him 1000 Deutschmark. Stark repeatedly asserted, and clearly believed, that it was only he who had saved German physics from extinction, and he enlarged on his achievement in a memoir, written in prison in 1945. An article in a journal, in which he professed to set the historical record straight, was edited to delete abusive comments about his enemies, deemed improper by the editor. Stark returned to physics and his theory of the doughnut-shaped atom, but seems to have failed to get his cogitations into print; physicists, he believed, now had insufficient grasp of fundamentals to follow his reasoning. His friendship with Lenard apparently ended late in the war. Lenard died in 1947, Stark ten years later.

Modern theoretical physics was still taught surreptitiously in many of the German universities during the ascendancy of *deutsche Physik*, and physicists continued to apply it in their research. Max von Laue made no attempt at concealment. He wrote to Einstein in America that when he expounded the Theory of Relativity in his student seminars he always reassured his audience that the original papers had all been written in Hebrew. Ludwig Prandtl, an applied mathematician and in his speciality of aerodynamics an international colossus, wrote to Reichsmarschall Göring that the interdiction of relativity and quantum mechanics as Jewish aberrations must stop before lasting damage was done to the country's capacity for technical progress—shades here of Kurchatov's *démarche* to Stalin—and Göring did indeed make a speech with a muted message that ideological dogma must not impede the course of science. It remained true that Einstein and other Jewish physicists could not be mentioned in publications, and an unworthy let-out was to impute their discoveries to others—relativity for instance to Hasenöhrl, a young physicist killed in the first World War, or to the great French mathematician, Henri Poincaré. A number of professorial chairs in physics were occupied by adherents of the *deutsche Physik*, but many more were filled by straw-men or left vacant; some chairs of theoretical physics were even abolished. In time, however, the physics community mounted a counterattack, which had the approval of Rust's ministry. Competent physicists managed to recover control of the University Teachers' Association, and a discussion on the subject of Aryan physics was arranged in Munich. Lenard and Stark did not attend and the defenders of their dogma were a poor lot, with only one physicist of any substance (Rudolf Tomaschek, professor at the Technical University of Dresden and a former student of Lenard) among their number. They were comprehensively vanquished in debate, and a statement was prepared, which was cautiously critical of the *deutsche Physik.* This was in practice the end, and with the appointment of a powerful and pragmatic industrial scientist to the presidency of the German Physical Society, normality was largely restored.

A deutsche Chemie

As in the Soviet Union, the political activists among the Third Reich's chemists found it difficult to formulate an ideological framework for their science. A movement was initiated, but, being so evidently absurd, failed to attract even one name of any consequence. There was for instance an attempt (like that in biology, as related above) to rid the technical language of foreign roots. *Absorption* (the same word in German as in English) was to be replaced by *Verschluckung*, or swallowing; *Elastizität*, elasticity, by *Dehnbarkeit*, or stretchability. *Maximum* and *minimum* were to be replaced by highest-value and lowest-value—*Höchstwert* and *Niedrigstwert.* Choice suggestions were *Haarrörhchenkraft*, hair-tubelet-power, for *Kapillarität* (capillarity), *Kleinsehwerkzeug*, or small-seeing-tool for *Mikroskop*, and, most outlandish of all, *Scheidekunst*—separation craft—for *Chemie.* Even the editors who published such glossaries in their journals seemed embarrassed, and the plan quietly died.

A number of practising Nazis in the community of chemists also tried to erect some cult figures from the past to serve as nationalistic exemplars for their *confrères.* One such was Theophrastus Bombastus von Hohenheim, more commonly known as Paracelsus, whose fourth centenary was celebrated in 1941. He was proclaimed a model anti-Semite (as indeed he was) and an early advocate of the *Blut und Boden* philosophy of the Nazi movement. The following year saw the centenary of Julius Robert Mayer, a pioneer of thermodynamics, given insufficient credit in his lifetime by reason of a conspiracy, so his admirers believed, by the British and French. Walter Gerlach, a respected physicist who should have known better, indicated in his address at the memorial symposium that Mayer's law of energy conservation expressed the principle that the self must be subjugated to the good of the *Volk*, in accordance with National Socialist teaching. Alwin Mittasch, a physical chemist, and an authority on catalysis, averred that Mayer's energy law applied equally to culture, that *Kraft*, or thermodynamic power, which stemmed from will, could be identified with *Volkskraft*.

There was more such blather. A small group of university chemists strove to define a *deutsche Chemie*, most prominent among them a professor of organic chemistry at the University of Leipzig, Conrad Weygand. German chemists, he urged, must shun the specious chemistry of the French and Anglo-Saxons, with its 'mechanical, materialistic, realism', deriving from a Cartesian view of the world and from the empiricism of Newton and Boyle (no hint here that it must be *anschaulich*). German chemistry must focus on matter (*Lehre vom Stoff*—doctrine of matter) and form. Form indeed was Weygand's preoccupation, for his research was concerned with the shape of chemical compounds and especially of their crystals. Weygand was undoubtedly a crank and his efforts were generally disregarded, as was the *deutsche Chemie* from beginning to end. A fervent patriot, Weygand met his end defending his city against the Russian army in the last days of the war.

Anti-Semitism and mathematics

Mathematics gave more opportunity for ideological casuistry, and the idea of a *deutsche Mathematik* took possession of several mathematicians of note, chief of whom was Ludwig Bieberbach. Mathematics in fact had long been the plaything of amateurs of racial psychology. In the first three decades of the twentieth century there were two great centres of mathematics in Germany, pre-eminent indeed in the world. The first was Berlin, but the University of Göttingen had come to rival and even surpass it under the leadership of the brilliant and domineering Felix Klein. To mathematicians the

Ludwig Bieberbach

rollcall of names of the members of the Mathematical Institute in Göttingen inspires awe: Klein, Hilbert, Minkowski (who laid the mathematical foundations for Einstein's relativity theory), Landau, Runge, Carathéodory, Weyl, von Kármán, Courant, Zermelo, Toeplitz, all these (and many more) were at various times there, some for most of their careers. An unusually large number were Jews, and indeed after it became known that Klein, in the course of a serious operation, had received a transfusion of blood donated by Richard Courant, the joke ran that the only Aryan mathematician in Göttingen had Jewish blood in his veins.

Late in his career Klein delivered, and published, a series of lectures on the recent history of mathematics, in which he developed a bizarre theory of racially-determined intellectual traits. In essence the Jews, in common with the Latins, were distinguished by a capacity for rapid and accurate analytical reasoning, while the Teutons supplied the qualities of imagination, ability to visualise in three dimensions, and intuitive grasp of intellectual truths. Of the English Jewish mathematician J.J. Sylvester (a true original with an exceptionally wide-ranging mind, who expatiated on such topics as the laws of verse), Klein wrote: 'In the brilliance and versatility of his spirit he was a true representative of his race'. Klein adverted to the influx of Jewish talent into mathematics, following the emancipation of the Jews in Prussia early in the nineteenth century. Together with the French immigration, this had enriched mathematics immeasurably. Klein saw this fusion of traditions as part of a general beneficent multiculturalism, to which he gave the unfortunate name of 'racial infiltration', and thus put a weapon in the hands of the National Socialist philosophers two decades on.

Others had entertained similar notions of racial types. The great mathematician Adolf Weierstrass had written to his protégée, the largely self-taught young Russian, Sonya Kovalevsky,[6] that his coeval, Leopold Kronecker, a cantankerous professor in Berlin, displayed the defects found in many highly intelligent people, 'especially those of Jewish origin', namely a lack of fantasy or intuition, for 'a mathematician who is not somewhat of a poet can never be a complete mathematician'.

By a curious paradox Felix Klein became a target for racial attack. It was alleged after his death in 1925 that his ancestry was Jewish, which was untrue, but more especially that he had allowed his institute at Göttingen (a city that was from the outset a stronghold of fascism) to be taken over by Jews—it was later described as the centre of 'the Jewish conspiracy' to dominate physics and mathematics. The accusation ran that Klein had used his influence with the reforming education minister, Friedrich Althoff (a fervent promoter of science in the universities and the founder of the Kaiser-Wilhelm Gesellschaft), to colonise other academic centres with his (too often Jewish) associates; and that he had sullied the purity of his subject by links with industry—all this besides being himself a representative of mathematics of the Jewish style.

The theory of racial stereotypes was developed by the camp-followers in the NSDAP, such as a psychologist, Erich Jaensch. He called the two types of mind the J-type and the S-type. Those in the latter category, to which the Jews belonged, were capable only of abstract, analytical and algorithmic analysis; they were rapid calculators (like their usurious brethren) and had keen memories. The S-type also included, rather oddly, the French mathematicians (all of them in the tradition of the pernicious Enlightenment) from Descartes to Cauchy and Poincaré. The fortunate possessors of the J-type mind were imaginative, intuitive, given to a synthetic approach, and inspired by nature. The leading champion of these grotesque ideas was a mathematician of high accomplishments, a former student of Felix Klein's and professor at Berlin University, Ludwig Bieberbach.

Bieberbach took exception to the 'modernist' trend in mathematics, which sought to develop all branches of the subject from a single set of axioms. The keynote was analytical rigour, and the high priest of the movement was Klein's successor as head of mathematics at the University of Göttingen, the great David Hilbert. Bieberbach insisted that intuition, geometrical vision and common-sense, not analytical formalism, must be the guide to mathematical truth, and in effect reiterated the demand by Lenard and his followers in physics for *Anschaulichkeit.* Bieberbach began to vociferate these precepts in 1930. He had crossed swords with Hilbert two years before over an international mathematics congress in Bologna. Since the Great War, European mathematicians—like the natural scientists—had excluded Germans from their meetings, but in 1928 the Italians decided that the time had come to forgive, and issued a round of invitations. Bieberbach and his ally, L.E.J. Brouwer, a fervid Dutch supporter of the NSDAP, wrote to all their *confrères* urging them to refuse the invitation and boycott the meeting. Hilbert declared that this attitude was foolish and destructive and himself went to Bologna at the head of a large German delegation. He also removed Bieberbach from the editorial board of the leading mathematical journal, the *Mathematische Annalen.*

Hilbert loathed the Nazis; he had blocked the appointment of Johannes Stark to a chair of physics at Göttingen, on grounds of his racial bigotry and political extremism, and he fought to prevent the dismissal of the Jews from the faculty in 1933. (When the Reich Education Minister, Rust, later visited the university and asked the ageing and dispirited Hilbert how mathematics was progressing in Göttingen, now that the baneful Jewish element had been extirpated, Hilbert replied: 'Mathematics in Göttingen? There isn't any left.')

In 1933 Bieberbach became an active Nazi, and aired his racial theories in a series of articles, extolling the intuitive style of mathematics as not only *anschaulich* (a concept eradicated during the previous century from the language of mathematics, in recognition of the often non-intuitive nature of mathematical thought) but also *völkisch*. It stood at the opposite pole to the sophistry that typified S-type mathematics, used by its adherents merely as a path to personal advancement. Bieberbach applauded the boycott by the Göttingen students of the lectures by the Jewish Edmund Landau, partly on the grounds that he had had the effrontery to express pi in terms of an infinite series—a typically Jewish subterfuge. G.H. Hardy, writing in *Nature* about Bieberbach's vapourings, observed that in times of political excitement one should make allowances for rash statements: 'Anxiety for one's own position, dread of falling behind the rising torrent of folly, determination at all costs not to be outdone, may be natural if not particularly heroic excuses. Prof. Bieberbach's reputation excludes such explanations of his utterances; and I find myself driven to the more uncharitable conclusion that he really believes them true.'

Bieberbach gained support from some of the younger academic aspirants, who discerned in his *deutsche Mathematik* an opportunity to leap aboard the accelerating National Socialist juggernaut; but he found little favour with established mathematicians, who saw in it the despoliation of their beloved subject. Even among the students, schizoid tendencies revealed themselves. A notable case was an outstanding young mathematician, Oscar Teichmüller; this Nazi activist and SA member cared nothing for J-types and S-types and continued to attend Landau's advanced seminars. He assured Landau of his personal and professional esteem, yet fully supported the boycott of the undergraduate lecture course, for nothing must be allowed to impede the campaign of racial purification. Teichmüller left his mark on his subject with a much-admired piece of work. When the war came, Bieberbach found him a position in a code-breaking unit in Berlin. But in 1943 Teichmüller apparently volunteered for service on the Russian front and was killed. Other Nazi students also continued to attend seminars by Landau and by another great Jewish Göttingen mathematician, Emmy Noether (held in her flat), before she left the country for the United States.

Bieberbach had in truth overestimated the party's interest in the purity of mathematics. The Ministry did not greatly exert itself to support his intrigues, aimed at taking control of the principal professional organisation of German mathematicians. He overreached himself in an exchange with the Danish mathematician, Harald Bohr (brother of Niels), who had denounced the German policy of expelling Jews from the faculties. Bieberbach published an open letter in a journal of which he was an editor without the consent of the editorial board. This allowed his co-editors, while conceding that Bohr's criticism was an attack on the National Socialist state, to censure their errant colleague for acting *ultra vires.* There was a further exchange in 1934 between Bieberbach and an exiled statistician, Emil Julius Gumbel, who had taught at Heidelberg and was now in Paris. Gumbel had excoriated Bieberbach for reducing a noble science to a wrangle over spurious racial differences. To Bieberbach's abusive reply, Gumbel published a rejoinder, ending with the statement that he was resigning from the German Mathematical Association, because that body 'in contravention of its statutes has offered no resistance to the destruction of German science'. Bieberbach's riposte read: 'As an answer to the pleasing letter of Herr Gumbel: Götz von Berlichingen. What do *you* know about *German* science? Bieberbach'. (The allusion to the drama by Goethe is a coded reiteration of the injunction, *Leck mich am Arsch*—lick my arse—to Germans a familiar verbal two-finger gesture.) Gumbel had the last word: 'I acknowledge receipt of your unpleasing letter and in the interests of the reputation of science, regret I am unable to answer in the same neo-German swineherd tone.'

Bieberbach attempted a *coup d'état* at a meeting of the German Mathematical Association, in conformity with the Nazi party's official *Führerprinzip*, entering the hall at the head of a mob of students, many in uniform. His demand for a change in the statutes and his bid for the chairmanship of the association were both heavily defeated in the voting. Bieberbach thereupon appealed to his principal supporter in the Reich Education Ministry; Theodor Vahlen, formerly a university professor and an applied mathematician, specialising in ballistics, had been a party member since the inception of the NSDAP and Gauleiter in Pomerania. Vahlen attempted to intervene on Bieberbach's behalf and succeeded in having the chairman of the association deposed. But Bieberbach was clearly unacceptable to the membership and a politically dependable compromise candidate was put forward and elected. This new chairman identified mathematics with the *Blut und Boden* movement, pronounced it a central component in the unity of body, mind and spirit, for which *völkisch* education must perpetually strive, and prepared a guide for mathematics teachers, in which such questions for the young featured as this: 'It costs 6 million Reichsmark to build a lunatic asylum. How many housing settlements at 16,000 Reichsmark each could be built with this sum?' Bieberbach withdrew from the fray and returned to Berlin University as Dean; Vahlen also granted him a handsome sum to start a new journal for *deutsche Mathematik*, which achieved a considerable circulation in the years before the war.

A curious case was that of the brilliant Pascual Jordan, a mathematical prodigy, who with Heisenberg and Max Born had participated in the great work which laid the formalistic foundations of quantum mechanics, known as the *Drei-Männer-Arbeit*. Jordan was an introvert, cursed with an incapacitating stammer, who had an interest in psychology and parapsychology. He was an early enthusiast for National Socialism and a member of the party and of the SA. His espousal of quantum mechanics brought him into conflict with Lenard and Stark, whose purblind *deutsche Physik* he deplored. He developed instead his own muddled conflation of the new physics with positivist philosophy and revolutionary politics. Physics, he asserted, encapsulated the principles of the Nazi movement, with its links to nature and its will to conquer. He also cleaved to the belief that physics must serve the interests of the state by developing technology for war. Jordan engaged in a series of lacerating polemics with representatives of the prevailing philosophical orthodoxy, which strongly opposed 'Jewish' positivism. Although he had earlier shown no inclination to anti-Semitism and some of his close associates, such as Max Born, were Jews, Jordan eventually got carried away by the tide of bigotry. He became overtly anti-Semitic, denouncing his dialectical adversaries as apologists for Jewish thought, and implied that Einstein's work was little more than an elaboration of Poincaré's. As the tide of the war turned against Germany, Jordan became noticeably more accommodating in his views. After the war he was arraigned before a denazification court, where he vehemently denied his recent utterances and claimed to have made the heroic decision to remain in Germany, rather than emigrating, so as to defend his subject and the good name of Jewish physicists against the incursions of the *deutsche Physik*. Jordan was shameless enough to ask his former friend, Max Born, to testify on his behalf. Born simply sent a list of family members and friends who had perished in the death camps. Yet Jordan was cleared of any serious misdemeanours, and unlike Bieberbach, was reinstated in the academic system and appointed professor at the University of Hamburg. He died in 1981.

The mathematicians for the most part behaved somewhat more creditably in the face of government bullying than other groups. The president of the professional association of applied mathematicians, who had acceded to his office because Jewish committee members had been forced out, stated that the only honourable course, if the expulsion of Jewish members could not be averted, would be to dissolve the society. The leading applied mathematician, Ludwig Prandtl, disagreed and petitioned that Jewish members be retained so that the international standing of the society not be damaged. He was in the end defeated: the Jewish members were expelled for supposed arrears of membership dues.

The consequences of the Nazi incursion into science

The Third Reich did not last long enough to cripple science utterly. The damage to biology was the worst, so much of it having fallen into the hands of the racial ideologues. Many established scholars, who had enjoyed a high reputation, participated with enthusiasm in this odious perversion of their science. A figure who stands out is Konrad Lorenz, ethologist and a later Nobel Laureate, whose writings before the war make a mockery of his self-righteous attempts to exculpate himself after it. The biology of Lorenz, Lehmann, Lenz and so many others meets the criteria of pathological science (Chapter 3), for it was based on a mirage. Without the political thrust of Nazi ideology it could scarcely have thrived, except as a marginal and despised activity of a few crankish outsiders, but here it was the academic establishment, the élite of the universities, who bore it aloft. This is not to say that good biology and medicine simply vanished after the National Socialists seized power. Indeed, as Robert N. Proctor shows in his absorbing book, *The Nazi War on Cancer*, the Party's preoccupation with the health and fitness of its citizens, so that they might the better serve the interests of the state, led to a number of prescient and enlightened measures. Tobacco was recognised as a poison and specifically as a carcinogen. In perceiving that industrial products—especially pesticides, asbestos and food colourings—were hazards to health the German epidemiologists were well ahead of their contemporaries in the Western countries. In part these concerns stemmed from the romantic view of the Nazi ideologues that civilisation had alienated the urban population from nature and had fostered an aberrant and unhealthy life-style. And so a campaign was instigated to promote physical fitness and a frugal diet and to discourage smoking and drinking. The Führer himself, as a vegetarian and teetotaller, who banned smoking in his presence, exemplified these virtues. The Party writ on such matters was in the event weakened by the power of industrial interests. In time also the attitude of the state came to resemble that of the society in Samuel Butler's *Erehwon*: sickness became a vice, a failure of patriotic will, while the chronically or genetically afflicted were, as we have seen, condemned as a burden to be cast off. As Robert Proctor puts it: 'The war on disease turned into war on the diseased'.

In physics and mathematics the consequences of the ideological aberrations were relatively slight—trivial, at all events, compared to the lasting damage done by the loss of the Jews and political subversives. The emigré physicists and mathematicians formed the backbone of the Manhattan project. The creation of the atom bomb was a blow to the self-esteem of their German contemporaries from which they never recovered. The Farm Hall transcripts—the bugged conversations of the German nuclear scientists captured by the Western allies and confined in a country house (Farm Hall) near Cambridge—reveal their incredulity and shock when the Hiroshima bomb exploded and they realised how comprehensively their own efforts had been eclipsed. The *deutsche Physik* and the *deutsche Mathematik* did not win adherents among any of the major figures in either field, except for the somewhat unhinged Lenard and Stark in physics and Bieberbach, Brouwer and Jordan in mathematics. The functionaries of the Third Reich recognised fairly quickly that these were not the true representatives of their communities, and that the measures they advocated would not assist the progress of science or the war effort. Even the powerful Himmler was marginalised where science was concerned, and, for the most part, genuine research prevailed. The damage to education was much more serious, because those who ran teaching departments in universities and the teachers in the high-schools were most often saturated with Aryan ideology. Thus students were deflected from the study of physics and mathematics, sensing probably that what they were being taught was deviant science. Physics teaching was completely distorted by the elimination of the modern theoretical foundations of the subject. In mathematics, 26 out of 95 university professors had been dismissed and nearly half of all lecturers. Many assistants and research students were expelled, or despaired and gave up. In Göttingen, for instance, that Mecca of mathematics, the number of mathematics students fell from more than 400 in 1932 to 37 in 1939. Germany has not recaptured anything like its pre-eminence in the fields of scholarship in which it could once take such pride, even to this day.

Notes

1 There was indeed only one recorded instance of a principled refusal to step into the shoes of a dismissed Jew in a biological or medical faculty: Otto Krayer in 1933 rejected the call to a chair at the University of Düsseldorf and wrote to the Minister for Science to protest against the injustices to Jewish colleagues. Krayer was forthwith cast out from the university and found his way to the United States. A unique case was that of Curt Kosswig, of the Technical University of Braunschweig, a member of the party and of the SS: he was a geneticist working on racial characteristics, who refused to endorse the policy of racial discrimination. He supported Jewish colleagues and, under pressure from his superiors, resigned his commission in the SS. He escaped from Germany just in time to avoid arrest. One half-Jew, interestingly, was not touched: Otto Warburg was perhaps the most prominent name in biochemistry in the country, and he retained, until the Russian army arrived in 1945, his position as head of a Kaiser-Wilhelm institute in Berlin, and even his stable of thoroughbreds, on which he liked to exercise (having been a cavalry officer in the First World War). Warburg was investigated, reclassified a quarter-Jew, and

was thus allowed to slip through the racial net. The reason was said to have been his claim that he was onto a cure for cancer, a disease of which Hitler was known to have a morbid dread.

2 Goudsmit was a Dutch-American physicist who led the Alsos mission, which followed the advancing Western armies through Germany in 1945 to assess the contribution of science to the war machine, and especially the state of research on the atomic bomb. *Alsos*, Greek for a grove, took its name from the administrative head of the Manhattan Project, General Leslie Groves. Goudsmit, a Jew, whose family had perished during the German occupation of Holland, was not sympathetic to the German scientists. He also found many targets for justified mockery.

3 Atoms absorb and emit light in the form of sharp lines at defined wavelengths. For instance, the familiar yellow light of a sodium lamp is due to a pair of lines at a wavelength of 589 nanometres (589 billionths of a metre), while the bluish-green light of a mercury lamp is due to a line at 546 nanometres. The visible part of the electromagnetic spectrum extends from about 420 to 700 nanometres, with violet at the short- and red at the long-wavelength end. When a magnetic field is applied to the atoms emitting the light, any line is split into two lines, very close together. This effect was predicted and was of great importance in validating the theory of atomic spectra. Canal rays, the subject of Stark's other discovery, consist of positively charged gaseous ions (see p. 75), which can be seen streaming through holes (canals) in the cathode of a vacuum tube containing gas at low pressure. The Doppler shift is a change in oscillation frequency of light, or of waves of any kind, receding from, or approaching the observer. It is most familiar as the changing pitch (frequency of sound waves) of an approaching or receding train whistle. Stark tried to interpret his observations in terms of special relativity theory and quantum theory, both of which he vehemently rejected in later years.

4 Haber, a baptised Jew, had kept Germany in the Great War with the Haber–Bosch nitrate process, which allowed the manufacture of explosives to continue after the Atlantic blockade had put a stop to imports of 'Chile saltpetre' (potassium nitrate). Patriotic to a fault, Haber had risked his career and his freedom by developing the German gas warfare programme and supervising its implementation on the Western and Eastern fronts. His wife, herself an organic chemist, had tried to dissuade him from this course, and when he would not listen to her pleas, she committed suicide; Haber left the next day for the front. After the war an attempt was made to arraign him as a war criminal. Later he initiated a scheme to pay off his country's war debt by extracting gold from sea-water, but this proved to be uneconomical. Although he won the Nobel Prize and became director of a Kaiser-Wilhelm Institute, Haber was never granted a professorial chair. One of his colleagues offered the following explanation: 'Before thirty-five he was too young, after forty-five he was too old, and in between he was a Jew.' When he was forced to resign and left Germany in 1934, Haber was a broken man: his services to his country counted for nothing and as an impoverished exile his self-respect was shattered. As a last resort he attempted to recover his Jewish roots and approached Chaim Weizmann, whose overtures he had earlier rejected, with a view to joining the Weizmann Institute in Palestine, but died before anything came of the plan.

5 The Uncertainty Principle states that the accuracy with which the momentum of a particle (in effect its speed) can be defined is coupled to the accuracy with which its position is known. The product of the two uncertainties, in momentum and in position, cannot be smaller than a fixed value, which is related to one of the constants of nature, Planck's constant. This carries with it the corollary that if one tries to observe an electron, say, by any physical method, the act of observation will change the electron's momentum. This sets limits on the power of experiments to study

the properties of fundamental particles and undermines in general the nexus between cause and effect—a profoundly unsettling conclusion to many physicists (including Einstein) and even philosophers. The Uncertainty Principle has been annexed by novelists and playwrights for their own purposes, to which it is seldom apt. Michael Frayn's play *Copenhagen*, produced at the National Theatre in London in 1998, deals with Heisenberg's crass action in visiting the intensly patriotic Bohr in German-occupied Copenhagen on a government-sponsored trip, and catches the flavour of the scientific, as well as personal, interaction between them.

6 She was born into the minor nobility of Russia, the daughter of a general. When she was a little girl the family home was redecorated, and her nursery was temporarily papered with pages from the general's mathematics books from his student days at the military academy. Sonya, in her cot, read the walls and became fascinated with their meaning. She trained herself as a mathematician and sent her work to the ageing bachelor and misogynist, Weierstrass, who recognised an exceptional talent and invited her to work with him. She later moved to Sweden, married, and died young, her last words, 'too much happiness'.

CHAPTER 11

Nature Nurtured: the Rise and Fall of Eugenics

The birth of eugenics

Eugenics, literally good birth, entered the language in the nineteenth century, with the dawn of genetics as a quantitative science. August Weismann in Germany had developed the germ plasm theory: the hereditary material of the offspring was carried in the parents' germ cells (p. 64), which were shielded from external influences, so that no environmental pressures could impose themselves on the genetic material. The idea that all human attributes, physical, intellectual and moral, were entirely, or at least primarily, heritable exerted a powerful appeal to many intelligent people, despite the weakness of the argument. As in other instances of pathological science, the flaws in the evidence on which the theory was based were overwhelmed by the wish that it should be true. Those who upheld the universal dominance of nature over nurture were, or felt themselves to be, intelligent, moral and free of the biologically deleterious traits that they eagerly sought out in others. In many cases eugenic doctrine was allied to a belief in the superiority of one race over all others and the conviction that its genetic purity must be maintained and protected. Certainly not all the early adherents of eugenics were malign, and many were dismayed by the consequences when eugenic measures were put into practice. Enthusiasts for eugenics ranged in their political allegiance from the extreme left (Prince Kropotkin was one) to the fascist right.

The story began with the wayward genius, Sir Francis Galton, and his motto, 'measure, measure'. Galton had tried to quantify all

Francis Galton, aged 42, in 1864. Wellcome Library, London.

manner of attributes of people, animals and plants, from weights and dimensions to physical beauty in women (the comeliest were in London, the most ill-favoured in Aberdeen). He had grasped the power of statistical analysis and in particular of the Gaussian, or normal, distribution. The Gaussian distribution, which takes its name from the great German mathematician, Friedrich Gauss, is now more popularly known as the bell curve. It applies to all statistically determined quantities, for instance how often a coin will come down heads when tossed a thousand times, or the heights of the men or the women in a population sample. Galton recognised that the essence of the Gaussian curve is that it is symmetrical about its

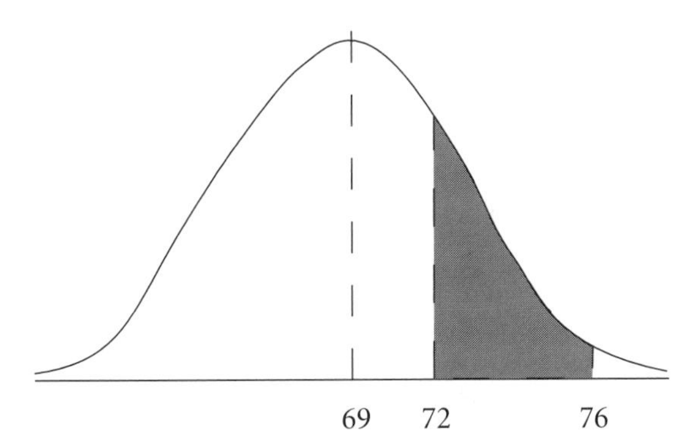

The Gaussian, or normal, distribution, also called the bell curve. This is the typical distribution of biological attributes in a population. The horizontal axis here is the height in inches of the men in a population and the height of the curve above the axis at any point measures the number (or fraction) of men with that height. Thus the average height in this pouplation is 69 in. The shaded area under the curve is the number (or fraction) of men with heights between 72 and 76 in. The fraction of men shorter than the average by the same amount (that is between 62 and 66 in) will be the same. Note that the number of men with heights above 76 in (or below 62 in) will be extremely small. (Under some circumstances asymmetrical, or *skewed*, distributions do occur).

maximum. So if, as in the illustration, the average height of the men in a large sample is 69 inches, the curve reveals that the number of men in this sample who are three inches taller than that will be the same as the number of men three inches shorter. Put another way, the probability that a man pulled at random out of the crowd is 66 inches in height will be the same as the probability that his height is 72 inches. The curve further tells us that the overwhelming proportion of men in the sample will be between say 62 inches and 76 inches. The little, roughly triangular, segments of the curve at either end will contain the very small proportion of exceptionally short or tall specimens. Why should such attributes as intelligence then not also conform to the same kind of statistics as height and weight? And if intelligence was heritable, should one not further expect the union of intelligent parents to produce intelligent offspring, just as the children of tall parents will tend to be tall?

Galton examined the heritability of intellectual performance in his famous book, *Hereditary Genius*, first published in 1869. He barely considered the possibility that a child growing up in a home in which a high value is placed on educational achievement, and with the stimulus that books and the parents' conversation will afford, might well do better than the same child growing up in deprivation. But Galton did recognise one limitation on the gift of the parental genes, which he called 'regression to the mean': a child of two very tall parents, for instance, is statistically likely to attain a height closer to the average in the population than to grow taller, or even as tall as, the stature of the parents might predict. All the more important therefore, if one wished to breed a hyperintelligent population, to select parents of the very highest intelligence, lying at the outermost edge of the Gaussian curve. Galton believed this could be done, and to encapsulate the principle he coined the word 'eugenics'. He urged the government to do all it could to promote genetically advantageous unions by offering various incentives, including free weddings in Westminster Abbey.

Eugenics and politics in Europe and America

Galton's cousin, Charles Darwin, was sceptical (though later persuaded), but these early eugenic concepts were generally received with interest. There was already apprehension about the rise of crime, alcoholism and venereal disease, especially among what were referred to as the lower orders, and these dispositions were widely seen as familial, for God had ordained that in the well-ordered Victorian society all men received their proper deserts. As the popular hymn had it:

> The rich man in his castle,
> The poor man at his gate,
> God made them high or lowly,
> And ordered their estate.

Across the Atlantic there were similar concerns, and Herbert Spencer's social Darwinism had already found an enthusiastic following. In 1877 Richard Dugdale published a study of the Jukes family from New York state. This brood of criminals, prostitutes, pimps and alcoholics could be traced through seven generations to a founder couple. How much pain and expenditure would have been saved, Dugdale asked, if they had been sterilised when the moral deficiencies of their deplorable dynasty became apparent?

The high-priest of eugenics for the next generation was Galton's young disciple, Karl Pearson. A cold, rather forbidding personality, he came of Quaker stock and had socialist and utilitarian convictions, and it was perhaps no coincidence that he found an academic home in the cradle of utilitarian philosophy and free thought, University College, London. Pearson had studied mathematics at Cambridge and adhered to the Galtonian doctrine that just about everything in biology was open to quantitative study. With the spread of the eugenic gospel, benefactors came forward to help establish the subject as an academic discipline. Pearson was soon presiding over a new institution of his own, the Biometric Laboratory, and, together with his colleague, Walter Weldon, another mathematically inclined biologist, and the ageing Galton, he founded a journal, *Biometrika.* The laboratory was enlarged after Galton's death, and became the Galton Laboratory for National Eugenics, with Pearson the first occupant of the Galton Chair of Eugenics.

Now began an extensive programme of data-gathering on the inheritance of many traits, from tuberculosis and deaf-mutism to alcoholism and, of course, intelligence. The heritability of intelligence was studied with the aid of a large sample of schoolchildren. Its measure was the opinion of the teacher. It emerged that by this simple criterion intelligence was indeed a familial attribute, with, according to Pearson, a correlation within the family group as strong as that for bodily dimensions. This led Pearson to the view that, since it was predominantly the intelligent middle classes who limited the sizes of their families, society was in peril of a swift intellectual decline, which means must be found to prevent. Such issues were debated in the now flourishing Eugenics Society (Major Leonard Darwin, son of Charles, president). Not all social reformers were overjoyed to be told that nothing could be done to improve the mental performance of the underprivileged, who were ineluctably condemned by their genes. Nor did all doctors welcome Pearson's assertion that tuberculosis was an inherited disease, for what virtue then remained in public health measures?

Many well-known public figures were drawn to eugenics in Britain, but it was in the United States that the movement attracted the most fervent support. The impetus came chiefly from the growing influx of immigrants then arriving from Ireland, Italy and Eastern Europe and widely perceived as riffraff, 'the sweepings of gaols and asylums', in the words of the racist author Madison Grant. At the head of the eugenic crusade rode Charles Davenport, a Harvard biologist with an interest in mathematics, who in 1899 took up an appointment at the University of Chicago and there proceeded to teach the virtues of eugenics. On a visit to England he called on the aged Galton and conducted an animated correspondence with him during the great man's declining years.

But Chicago did not allow Davenport's ambitions their full scope and soon he became director of a summer experimental station at Cold Spring Harbor, not far from New York City on the northern tip of Long Island (where it thrives today in another guise—as one of the

Karl Pearson

Charles Davenport

great centres of molecular biology). There, with the support of the Carnegie Institution, a philanthropic organisation recently founded by the Scottish industrialist Andrew Carnegie, Davenport was enabled to realise his vision of a programme of research into human heredity. Davenport was evidently a man of considerable charm and eloquence, for in 1905 he persuaded the heiress Mary Harriman to disburse what at that time was a huge sum for a larger project, the construction and operation of a laboratory, to be known as the Eugenics Record Office. Davenport set in motion a campaign to gather and analyse vast amounts of data on the American population.

Davenport was deeply interested in racial differences, and seemed to accept all the clichés about the character traits of different nations: the Irish were drunk and feckless, the Slavs dimwitted, and so on. Aberrant behaviour, such as crime and alcoholism, was, Davenport believed, inherited and not the product of destitution. Prostitution arose not from want, but from 'innate eroticism'.

It was no coincidence that Davenport made friends with the German racist academic and avid supporter of the Nazi party, Eugen Fischer (Chapter 10). Davenport sent to Mussolini a report written by Fischer, regarding the urgent need for a eugenic policy if the looming threat of racial decline was to be averted. Fischer's speciality was the study of mixed-race offspring, and Davenport invited him in 1929 to chair a new commission on the subject under the aegis of the International Federation of Eugenic Organizations. Three years later Davenport nominated Fischer again, this time to succeed him as president of the Federation, although the invitation was declined.

At home, Davenport found a useful ally in Henry Goddard, who introduced into America the supposedly objective intelligence test, devised for the French government by Alfred Binet. This consisted of a series of simple questions, and was developed by Goddard and the psychologist, Lewis Terman, into the Stanford–Binet test. Mass-testing was practised on a captive population—recruits in the army: a large proportion were judged to be 'feeble-minded', especially among the black and Eastern European immigrant groups. The tests were also applied—in English—to new immigrants, who generally spoke no English, and produced the conclusion that up to half were feeble-minded. This alarming outcome hastened the introduction, at Goddard's urging, of the Immigration Act, which abruptly reduced immigration to a trickle.

Dysgenic families were produced as evidence for the inheritance of asocial behavioural traits, most famously the (pseudonymous) Kallikaks: Goddard published his study of these children—delinquent or 'feeble-minded'—of the illegitimate liaison of a respectable white father with a mentally defective barmaid. When the father later married a clean-living, God-fearing wife, wholesome offspring resulted. In Goddard's photographs the dysfunctional Kallikaks appeared lumpen or more often sinister, but it was later discovered that their features had been crudely doctored with black ink. (The view that a propensity towards criminal or other antisocial behaviour was accurately reflected in the physiognomy had assumed in such circles the status of a dogma: it dated back to the writings of Cesare Lombroso in Italy in the late nineteenth century. Lombroso claimed to have identified the distinguishing facial features of different classes of criminal—murderers, adulterers, pickpockets, prostitues, rapists and many more.)

Goddard and Terman (who coined the term Intelligence Quotient) were persuaded that intelligence was the product of a single gene[1] and was inherited in simple Mendelian fashion.[2] Goddard was troubled by the imprecision that attached to the designation 'feeble-minded'. He did not believe that there was a continuum of IQ in the population. There were the pathologically abnormal—the idiots and imbeciles, as defined by the clinicians—and there were the merely stupid. For these last Goddard invented a new word with a Greek stem: moron. Above them stood the labouring classes, slow-witted but sound, and above these the intelligent. Both the morons and the intelligent were the homozygous progeny of two homozygous or heterozygous[2] parents. Thus two morons could give rise only to moron progeny and two of the intelligent group only to intelligent children. But the union of two heterozygotes (labourers), each carrying one gene for intelligent and one for moronic character, would lead to an average of one intelligent offspring out of four, one moron, and two of the heterozygote labouring class. Goddard had no doubt that the three types could be easily identified, commonly at sight. (Women had the more acute perception, and it was two women assistants that he sent to examine the immigrants on Ellis Island.) Goddard's further insight was that morons possessed limited control over their biological and social urges and made up the bulk of the criminal classes and social misfits.

It was not long before a majority of the States of the Union introduced eugenic measures, mainly marriage laws, to exclude those with genetic diseases and the socially dysfunctional, and in some cases to prevent miscegenation. Worse, compulsory sterilisation became common. In the state of Indiana it had been legal since 1907, when many patients in mental institutions were forcibly sterilised for such lapses as persistent masturbation. In judging a landmark legal battle over the sterilisation of a child deemed to have inferior intelligence, the offspring of a supposedly mentally defective mother, the literary Justice Oliver Wendell Holmes declared: 'Three generations of imbeciles are enough.' It was his affirmed belief that 'it is better for the world if society can prevent those who are manifestly unfit from continuing their kind'. Some tens of thousands of sterilisations, by castration, vasectomy or irradiation, were performed in the United States. The Cambridge geneticist Reginald Punnett calculated that if there were indeed a recessive gene for feeble-mindedness, the American eugenics programme would result in a reduction from 3 per cent incidence in the population to 1 per cent in something over 8000 years.

The worst excesses occurred in Virginia, where whole families of the unemployed on welfare were rounded up, and many women and girls were sterilised. Large numbers of frightened citizens took to the hills to avoid this fate. In 1934, as the new Nazi regime in Germany was getting into its stride, Dr Joseph De Jarnette, one of the most strident advocates of mass sterilisation in the State, complained that the programme in Virginia was not proceeding with enough dispatch: 'The Germans are beating us at our own game.'

Syphilis was another condition commonly adjudged to be genetically determined. It was regarded as a 'Negro disease', and in a poor rural county in Alabama studies on infected black men were pursued over a period of forty years. Many were given only placebos and the advance of the disease was recorded until the victims died.

Compulsory sterilisation continued until long after the Second World War and laws permitting it remained on the statute books in many States of the Union until well into the 1960s. Euthanasia for the feeble-minded was urged in some quarters, but was never (officially at least) introduced. Goddard himself had long before become convinced by genetic arguments that he had been wrong in his assessment of heritability of intelligence and made a public recantation. In the following decade, enthusiasts for eugenics were still advocating voluntary sterilisation with pecuniary incentives. A vocal champion for a nationwide scheme in the 1970s was the renegade Nobel Laureate in Physics and apologist for white supremacy, William Shockley. (It was he who set up a sperm-bank, all samples to be donated by men of genius, such as himself, to improve the intellectual and moral stock of the American nation.) There were undoubtedly cases in which parents were tricked into allowing their children to be sterilised.

In the first two decades of the twentieth century, eugenic legislation was introduced in many countries, most ominously in Germany, but also in Canada and throughout Scandinavia. In Sweden, for example, an institute for Racial Biology was founded as early as 1921. Compulsory sterilisation became legal in 1934 and, over the next thirty years or more, no less than 1 per cent of the population—some 63 000 Swedes—were sterilised on grounds of racial or social undesirability. About 95 per cent of these were women—single mothers or those who had borne 'unruly' or 'feeble-minded' (including epileptic) children. Unless the subjects were judged incapable of understanding what was to be done to them, sterilisation was in principle voluntary, but in practice there were almost irresistible pressures on the victims—threats of perpetual detention, for instance—to consent. The under-educated community known as 'the travellers', who, unlike the gypsies, were not even a distinct ethnic group, were treated with especial severity. The law which permitted these iniquities was not repealed until 1976, the last institution to give it academic respectability, the Department of Race Genetics at the University of Uppsala, having closed the previous year.

In Britain, the birth-place of eugenics, the movement had many distinguished adherents, including George Bernard Shaw, H.G. Wells, Havelock Ellis, Sidney and Beatrice Webb and J.B.S. Haldane, but 'negative eugenics' had no supporters of note and even legislation to allow voluntary sterilisation was rejected. Indeed, a milder measure, the Mental Deficiency Bill, which had come before Parliament in 1912, and was strongly supported by the Eugenics Education Society, faced passionate opposition and was defeated. The bill proposed segregating adults certified by two physicians as mentally defective, to prevent them from procreating, though treating them humanely in all other respects. The measure was seen by its opponents as the thin end of a totalitarian wedge. Noting that the criteria for mental inadequancy had been formulated by the Royal College of Physicians, one of the bill's parliamentary opponents observed: 'If there is anybody I would less trust in a matter of personal liberty than a bureaucrat, it is an expert.'

Scientific opinion on heritability of asocial attributes remained divided, however. A decade later the editor of *Nature*, Sir Richard Gregory, could still write the following: '… a large proportion of the slum populations consists of … "morons"—that is of mental defectives of comparatively high grade. These people are lacking not only in intelligence but also in self-control, which is the basis of morality, and they reproduce recklessly'.

The fairly numerous advocates of mild 'positive' eugenics—mostly encouraging the middle-class intellectual élite to procreate more vigorously—fell away as events in Hitler's Germany unfolded. A letter to

Nature, urging extreme measures, in 1936 by the zoologist, Professor E.W. MacBride, a veritable dinosaur in his biological opinions (he was an unreconstructed Lamarckian and believer in spontaneous generation) drew a stinging response from the biochemist (and celebrated sinologist), Joseph Needham:

> the compulsory and *punitive* sterilisation of parents who 'have to resort to public assistance in order to support their children' is offered as a remedy. Are we to assume that E.W.M. includes shipowners, beet-sugar shareholders, and other persons receiving financial benefit other than wages from industries subsidised by the State, though privately owned, in this category? And can he even be serious in suggesting 'punishment' for the two million unemployed?. It is difficult to express the dismay experienced in seeing these doctrines, so dangerous to humanity, receiving the imprimatur of what is perhaps the most famous scientific weekly in the world.

The most authoritative and tenacious scientific opponent of eugenics was Lionel Penrose, who was later to occupy Pearson's chair at the Galton Laboratory (the word 'eugenics' was removed from its name at his behest) at University College, London. Penrose was a Quaker, a socialist and founder of a distinguished scientific dynasty. He was a geneticist with a particular interest in mental disorders. His objections to eugenic policies were both scientific and moral. A compassionate society must provide shelter and care for the genetically disadvantaged, rather than mutilate them. It was also evident to him (as to the great American geneticist, Thomas Hunt Morgan) that to regard mental defects, and for that matter intelligence, as simple Mendelian characteristics was absurd: they were multifactorial and influenced by external circumstances. More especially, such states as 'feeble-mindedness' were not rigorously defined and their identification relied on subjective assessments. Further, the efficacy of eugenic measures, such as sterilisation, in eliminating a given gene from the population had not been thought through; and as to the family studies, such as Dugdale's for instance on the notorious Jukes brood, the proportion of supposedly affected members was far higher than Mendelian inheritance allowed. In Britain, Penrose's view prevailed.

Eugenics in the Third Reich

In Germany the concept of race, as we have seen, had mythical power, and the necessity of guarding its purity and preventing its degradation by diseased alien stock was embedded in the teaching of Haeckel and his successors. The eugenic movement therefore gathered momentum early and became inextricably associated with the pseudo-academic discipline of *Rassenhygiene*. Whereas in Britain and the United States eugenicists were in general humanists or statisticians, the movement in Germany was dominated by doctors. The first professorial chair with that title was established at the University of Munich in 1923, and Fritz Lenz received the call. Lenz was from the outset a supporter of Hitler's NSDAP. He and two like-minded colleagues, Erwin Baur and the infamous Eugen Fischer, published what became the standard textbook of genetics and racial hygiene in the Third Reich.

Sterilisation of patients in mental hospitals was apparently already common, though illegal, in the 1920s. Much of the impetus was undoubtedly economic, for this was the time of crippling recession, but the concept of *lebensunwertes Leben*, or life unworthy of life, had already been articulated by the political right. (The expression came from the title of an influential book.) With Hitler's accession to power in 1933, things began to move rapidly. Legislation was introduced at the beginning of the following year for the elimination of genetically diseased offspring. Later the same year a law was passed to legalise sterilisation of dangerous and persistent criminal offenders. This relied, of course, on the axiom that criminality was an inherited trait.

The supposedly genetic conditions for which sterilisation now became mandatory were feeble-mindedness, schizophrenia, epilepsy, hereditary deaf-mutism, alcoholism, Huntington's disease, manic

Gesetz zur Verhütung erbkranken Nachwuchses

vom 14. Juli 1933

mit Auszug aus dem Gesetz gegen gefährliche Gewohnheitsverbrecher
und über Maßregeln der Sicherung und Besserung vom 24. Nov. 1933

Bearbeitet und erläutert von

Dr. med. Arthur Gütt
Ministerialdirektor
im Reichsministerium des Innern

Dr. med. Ernst Rüdin
o. ö. Professor für Psychiatrie an der Universität und Direktor
des Kaiser Wilhelm-Instituts für Genealogie und Demographie
der Deutschen Forschungsanstalt für Psychiatrie in München

Dr. jur. Falk Ruttke
Geschäftsführer des Reichsausschusses für Volksgesundheitsdienst
beim Reichsministerium des Innern

Mit Beiträgen:

Die Eingriffe zur Unfruchtbarmachung des Mannes
und zur Entmannung
von Geheimrat Prof. Dr. med. Erich Lexer, München

Die Eingriffe zur Unfruchtbarmachung der Frau
von Geheimrat Prof. Dr. med. Albert Döderlein, München

Mit 15 zum Teil farbigen Abbildungen

J. F. Lehmanns Verlag / München 1934

Title page of a book published in 1934 to expound the Law for the Prevention of Genetically Diseased Progeny, enacted in the previous year. It was mandatory for all German doctors to buy this book. The publisher, as usual, was J.F. Lehmann, owned by the brother of the racial biologist Ernst Lehmann.

depression and bodily malformations. Births of abnormal babies and of babies to parents with these defects were to be reported by doctors, midwives, community nurses, dentists and others. Many incompetent sterilisations and abortions were carried out and deaths were not uncommon, nor indeed were suicides. It was estimated that between the passing of the law and the outbreak of the Second World War something like 300 000 German citizens were sterilised. The policy was made more palatable by an extensive propaganda campaign,

Cartoon from the eugenic journal *Volk und Rasse* of 1933. The caption reads: Here you are bearing your part of the burden. A genetically diseased person costs, up to his 60th birthday, on average 50,000 Reichsmark.

emphasising the burden to the community and to families of supporting 'life unworthy of life'. Posters showed pictures of sweating workers bearing across their shoulders yokes on which perched deformed children. The propaganda was directed equally against those other social encumbrances, beggars, alcoholics and the habitually workshy. These were soon rounded up by the police and confined in camps, where they were employed in forced labour under inhumane conditions, from which many died.

In 1935 the marriage laws were added, forbidding those adjudged to carry undesirable genetic traits to marry. These traits were numerous and Josef Mengele's teacher and patron, Professor von Verschuer (p. 224), even pronounced tuberculosis to be hereditary (echoing the opinion of Karl Pearson). The next year, under conditions of close secrecy, killings got under way in hospitals and asylums, carried out by doctors, some of whom used injections, others slow starvation. The administrative director of each institution was given the responsibility of selection, in accordance with the *Führerprinzip*. The families were told death had resulted from a mishap or illness, such as pneumonia.

In September of 1939 the regime took the next logical step and passed a euthanasia law, to replace sterilisation. Separate administrative authorities supervised the murder of adults and of children. Doctors were issued questionnaires on all patients in existing institutions. The forms were processed by officials at the Interior Ministry in Berlin, who turned their thumbs up or down, according to their evaluation of the information. The victims were taken to killing centres, administered by the Euthanasia Office, known as Aktion T4, in Berlin. They were examined, rarely reprieved, stamped on their backs if they had gold teeth or fillings, and gassed. The gold was extracted and the bodies cremated. Often organs, especially brains, were first removed and sent to various university departments for study, in the expectation that their anatomy would reveal something of the cause of the victim's disability. The main beneficiaries were Professors Hallervorden of Berlin and Schneider in Heidelberg, who were punctilious in their expressions of gratitude for the 'rare and precious material'. Experiments on living subjects were not excluded. In all, not less than 70 000 patients in institutions were killed.

'Positive' eugenic measures were also tried, for Professor Lenz and other experts were of the opinion that only a minority of the German population were fit to bear children of a quality to maintain or improve the genetic constitution of the race. Notable was Himmler's project for breeding centres for the SS élite, launched in 1936. Obstetric facilities were made available for the pregnant women, mainly SS wives or mistresses, and families were looked after in these homes. Pedigrees were required of parents, who were rigorously screened for Aryan purity and freedom from undesirable familial traits. As far back as 1931, members of the SS wishing to marry had to submit their prospective brides for inspection by the SS Race Office. A premium was placed by the State on fecundity. (Mothers of four qualified for a bronze medal, for six children they received a silver and for eight a gold medal.) Unmarried women of suitable descent were offered impregnation by an SS man of their choice in person or by artificial means. The project was only a limited success and the institutions atrophied with the deterioration in conditions as the war progressed.

Eugenic nemesis in the Soviet Union

Eugenics aroused as much interest in pre- and post-revolutionary Russia as anywhere else. Both Marx and Engels had seen virtues in the concept. The Russian Eugenics Society was founded in 1920, along with two journals dedicated to the new discipline. The next year a Bureau of Eugenics was established in Leningrad under the leadership of a respected geneticist, Yuri Filipchenko. The most prominent academic associated with the subject was the country's leading human geneticist, Nikolai Koltsov, first president of the Russian Eugenics Society. Many early Bolsheviks embraced the vision of scientifically regulated social progress, and supported the evolution of a better society, assisted by a campaign of sterilisation of undesirables and mass insemination of the womenfolk with the sperm of selected males of exceptional ability. Such notions were repugnant to most biologists, however, Filipchenko and Koltsov among them.

It was on the whole political liberals who dominated the field, and they were backed by several prominent Western European and American eugenicists, sympathetic to the Soviet Union. The Austrian socialist, Paul Kammerer, Western fugleman of Lamarckian inheritance (Chapter 7), accepted a permanent position at the Soviet Academy of Sciences. He almost certainly planned to occupy himself with eugenic studies but, as already related, committed suicide shortly before he was due to take up his appointment.

Kammerer had written that Mendelian genetics made men 'slaves of the past', while Lamarckianism promoted them 'captains of the future'. But to biologists in general the Lamarckian doctrine was already discredited and the province of crackpots (the most alarming of whom, Lysenko and Michurin, were then as yet a mere cloud on the horizon, no bigger than a man's hand). In any event the first Five-Year Plan was launched in 1929, and with it came Stalin's 'Great Break'. Plans for technological control of society were suddenly seen as bourgeois and reactionary, if not fascist. The term 'biologism' was heard, and advocates of eugenics were branded 'Menshevising idealists' in the hastily revised biological section of the Soviet Encyclopaedia, always a thermometer of the political temperature. (The difficulty of predicting the past in the Soviet Union was noted in the West.) Wholesale dismissals and the disbandment of genetics laboratories followed, the chapters on human genetics disappeared from translations of Western textbooks and geneticists vanished into Siberian or Caucasian exile.

A year or two later the wind changed again, and genetics made a cautious return. H.J. Muller, the American communist geneticist, had committed himself to emigrating to the Soviet Union at a time that seemed to him propitious for the dissemination of his eugenic ideals. He arrived in 1932 to head a laboratory in Vavilov's new Institute of Genetics, was warmly welcomed as a highly desirable catch from both the scientific and political standpoints, and only a year later was elected a Corresponding Member of the Soviet Academy of Sciences. Muller's research centred on fruit-fly genetics, but he wrote popular articles elaborating socialist eugenic programmes. His views stemmed from a conviction that nearly all mutations, accumulated at random over the generations, were deleterious, often lethal. He referred to the sum of the individual's unfavourable traits as 'the genetic load'. Medical advances and improved living conditions would allow the unfit to survive and procreate. In Muller's apocalyptic vision a few generations would suffice to bring forth a population of monsters. And so he developed eugenic prescriptions for a good society, founded on a programme of artificial insemination with the sperm of a genetic élite. He enlarged on this in the book that so irked Stalin with the consequences already described (Chapter 9).

It was not until after Krushchev's overthrow and the fall of Lysenko that human geneticists emerged once more into the light, determined now to say what they had bottled up through the years of oppression. In the 1970s a furious debate erupted between the adherents of the nature and nurture schools, on ground long abandoned by the geneticists of the West. The postures of the adversaries on such matters as inheritance of criminality were informed more by political than scientific convictions. Personal animosities and the itch to settle old scores also prevailed and must have clouded the reason of those who had suffered so unjustly. That generation has now vanished and Russian biology has different problems to contend with.

The rise and fall of eugenics: a pathological science

The history of eugenics affords another measure of the degree to which opinions, which should be based on a sifting of hard-won scientific evidence, can in reality be moulded by political beliefs or social pressures, not to mention mere self-interest. Scientific principle under such constraints is apt to yield to the desire to be on the right side. Lenz and even Fischer in Nazi Germany are good examples. Lenz, before 1933, was philosemitic in his utterances, and Fischer in his early work placed the Jews well up the racial ladder. Both changed their views with indecent speed as the new political orthodoxy emerged (and back again with equal dispatch after the defeat of the Third Reich). Their textbooks of genetics first incorporated as scientific fact, and, the war over, expunged, the racial precepts enunciated by the ideologues of the Third Reich. Many of the doctors who implemented the sterilisation and euthanasia programmes with diligent brutality were described by their associates as kind and conscientious men, who were evidently able to delude themselves that they were doing right by society, according to scientific imperatives.

Today eugenics is (more or less) a dead letter, hastened to its end by the events in Nazi Germany. There was a brief and half-hearted resurgence a decade or two after the war ended, when fears arose that medical advances might negate the ill-consequences of some genetic disorders. Thus phenylketonuria is a hereditary failure of metabolism, previously fatal but now easily detected and rectified by a lifelong diet. Treatments for such conditions as sickle-cell disease are improving and patients are living longer and indeed bearing children. It used to be argued that such clinical intervention in the natural course of diseases would disastrously increase the incidence of the deleterious genes in the population, but it is now generally accepted that the effects are small.

Only in China, where the rights of the individual are regarded as of no account compared to the common good, is eugenics still in favour. A Eugenics Society has high party officials on its council and airs its concerns about the burden on society of children with such afflictions as Down's syndrome; the figure of 30 million genetically impaired Chinese citizens is often bruited. Marriages between partners carrying known deleterious traits are discouraged, and a law enacted in 1995 compels couples to be passed genetically fit for procreation by a doctor. Abortion or sterilisation is urged for preventing the birth of genetically handicapped children, though they are in principle voluntary. Some eugenicists have recommended the introduction of infanticide where such measures have failed, and a debate along these lines apparently continues.

Eugenics may be dead in the West, but its close relative, racism, still thrives. The debate about race and intelligence recurs every few years, even though the racial groups in question all have mixed gene pools, and the effects of culture and family background on intelligence tests remain imponderable. The preoccupation of German biologists, many of proven scientific competence, with racial purity throughout the first half of the twentieth century is now all but impossible to grasp. Julian Huxley made fun of it when he described the ideal Teuton in 1935 as being 'as blond as Hitler, as tall as Goebbels [who was short and club-footed], as slim as Goering [hugely corpulent], as dolichocephalic [long-skulled] as Rosenberg and as manly as Streicher'. Odd that such thoughts never disturbed the minds of the racial hygienists.

The broader debate about the relative influences of 'nature and nurture' on mental and behavioural characteristics continues yet. Scientific opinion is polarised according to political preference: the left is for nurture, the right for nature. The results of identical-twin studies swing one way and then the other. Sir Cyril Burt's work on the intelligence of identical twins, separated at birth, was discredited by evidences of fraud, although there have since been attempts, not altogether convincing, at rehabilitation. Recent studies seem to have tilted the balance back a little towards nature. The answer surely lies somewhere in between, but just where may never be fully resolved.

Notes

1 To support this extreme hereditarian view, Terman, with his colleague, Catherine Cox, undertook a study of 312 geniuses from history. The estimates of their IQ relied on information about their early development and ancestry, drawn from biographical sources. At the apex stood Goethe, with a prodigious IQ of 200 (average being by definition 100), closely followed into second

place by John Stuart Mill, weighing in at 190. Practically all the geniuses had IQ far above the ordinary, however. The credulity of Cox and Terman seems to us now fathomless, considering the exiguous nature of the information about the early years of long-dead historical figures, and the absurdity of placing a quantitative valuation on any of it. But the conclusions found their way into the textbooks of the day and were solemnly absorbed by a generation of students of education, sociology and psychology.

2. Mendelian inheritance takes its name from Johann Gregor Mendel, the monk whose studies on sweet pea plants in the garden of his monastry in Brno (then Brünn) in Bohemia uncovered the basic principles of heredity. As an illustration consider the inheritance of the human disease, sickle cell anaemia: in sexual reproduction one set of chromosmes bearing our genes comes from the father, a second from the mother. Our cells contain 23 such pairs. When the parental chromosomes recombine in the offspring they do so at random. Suppose now that the father has a gene on both pairs of one chromosome with the abnormality (mutation) that gives rise to an aberrant version of the oxygen-carrying protein, haemoglobin. His genetic make up in respect of haemoglobin can be written **SS** (for sickle). If the mother has only the normal gene on both chromosomes of the pair (**AA**) each child will inherit an **A** from the mother and an **S** from the father. They will all have the genotype, as it called, **AS**. But if both parents carry the abnormal gene on *one* of their chromosome pairs and thus have the genotype **AS** then a child can inherit the **A** or the **S** gene from either parent and will therefore have a 25% chance of turning out **AA** (fully normal), a 25% chance of being **SS** and a 50% chance of being **AS**. The **SS** condition results in severe, in earlier times lethal, disease (sickle cell anaemia), the **AS** in a barely pathological state (sickle cell trait). Individuals with the same gene on both chromosomes are said to be *homozygous* for normal or sickle cell haemoglobin and those with different genes (one **S**, one **A**) are *heterozygous*.

Envoi

What (if any) morals can one draw from the foregoing episodes of human folly and delusion? These stories are not about merely being wrong. That, after all, happens all the time in science, as in every other human endeavour, and often to the best of scientists—those who make the imaginative leaps into the unknown. A political commentator has asserted that the function of an expert is not to be right but to be wrong for more refined reasons; and the great Niels Bohr defined an expert as one who has made all the mistakes that can be made within his narrow field. It is confusion, rather than error, that impedes progress. It is not even a matter of being grossly and embarrassingly wrong (though this was certainly a component in the history of cold fusion and memory transfer, for example); that too can happen to anyone brave enough to take risks. It is a corollary of Murphy's Law (which, as originally enunciated, states that if anything can go wrong it will, with the rider, even if it can't it may) that the more foolish the action the more people will be watching. Since all scientists are well aware of this, mortifying lapses are generally treated with at least some degree of sympathy.

We are also not concerned here with irrationality; advances especially in theoretical physics, which deals in abstractions far removed from the world of sensory experience, often rest on imagination raised to a higher plane. The physicist and writer, Jeremy Bernstein, has given a memorable account of a clash between two of the titans of the subject, Niels Bohr and Wolfgang Pauli, after a lecture by the latter. Bernstein observed the two ageing mastodons, manoeuvring around the lectern in their effort to confront the audience, Bohr declaring the while that Pauli's new theory might appear crazy, but was not crazy *enough*, and Pauli riposting that is was quite satisfactorily crazy. (It was wrong; Pauli was in terminal intellectual and physical decline, and Freeman Dyson, watching from the hall, likened the scene to the death of some noble beast). But craziness—a total break with what the philosopher of science, Thomas Kuhn, termed the prevailing paradigm—was the essence of the greatest discoveries of twentieth-century physics, of quantum theory and of General Relativity.

Much has been made of Max Planck's famous observation that new theories gain acceptance not because the community is persuaded by force of argument, but rather by the passing of a generation of scientists and its replacement by its younger, more open-minded successors. But, in truth, what seems more remarkable in the light of history, is the speed with which the new intellectual order that resulted from the conceptual cataclysms *was* accepted by the great majority. In biology, the opposition to Darwin and Wallace came much less from within science than from outside. (The public view, initially at least, was probably best encapsulated by the remark overheard at the end of the famous Oxford debate on evolution in 1860; it was by a lady to her companion: 'Let us hope it is not true. But if it is true, let us hope it does not become generally known'.)

This is not to say that a change in accepted orthodoxy owes nothing to fashion or chance. James Clerk Maxwell put it like this: 'Once we believed in the corpuscular theory of light. Now we believe in the wave theory. And this is because all those who believed in the corpuscular theory have died'. (The truth, as we now know, is that light is both corpuscular and wavelike.) But this does not negate the central part that scepticism plays in the scientific process. A purported Kuhnian paradigm shift—the overthrow of an established view—has to pass through the fire of criticism, even animosity, before it can establish itself, otherwise credulity reigns and progress ceases.

How then are we to recognise what I have called (after Irving Langmuir) 'pathological science' and distinguish it from an authentic conceptual leap that transcends the wisdom of the day? An oft-cited example of a hypothesis that divided the experts and attained a moderate following, but was fiercely (and wrongly) denounced as nonsense by most of the established figures in the field, was Alfred Wegener's theory of continental drift. To geologists Wegener was something of an outsider, having been trained as a meteorologist. He was also a German, who had fought in the Great War, in which he was twice wounded. This probably contributed to the hostile reception that he received in the United States and Britain in the early 1920s (though he had in fact developed pacifist tendencies as a result of his harrowing experiences). His book, *The Creation of the Continents and the Oceans*, running into many revised editions, elaborated his theory that the continents had separated from a single primaeval land-mass, which he called Pangaea; it was much reviled, though welcomed by the palaeontologists, whose observations on the evolutionary similarity of fossils in the New and the Old World called for a vanished bridge between the two. It was only decades after Wegener's death (on an expedition to Greenland) that continental drift became the new orthodoxy and gave rise to the science of plate tectonics.

Wegener's book did not, however, develop the theory in an entirely convincing manner and contained assertions that were contrary to observation, as well as conspicuous lacunae. It was a long time before enough persuasive evidence accumulated to give the theory full credibility. It cannot, at the same time, be denied that much of the opposition that Wegener encountered was overwrought, often verging on the hysterical. Geological, like astronomical phenomena could not be tested experimentally, a circumstance that invited philosophical wrangling. Here, for instance, is the opinion proffered by the American geologist, Henry Fielding Reid: 'There have been many attempts to deduce the characteristics of the earth from a hypothesis, but they have all failed .... This [the theory of continental drift] is another of this same type. Science has developed by the painstaking comparison of observations and, through close induction, not by first guessing at the cause and then deducing the phenomena'. Reid was evidently one of the last followers of Francis Bacon. Most scientists, and for that matter philosophers of science, are, as we have seen, agreed that observations with no theoretical framework are futile.

It is perhaps evading the question to conclude that pathological science can always be recognised for what it is from the perspective of time. The abrupt disappearance of all mention of cold fusion or of memory transfer from the scientific literature (p. 80) meant that even their most fervent adherents eventually lost faith, as the evidence for the defence came to appear increasingly exiguous and that for the prosecution harder to contest. But we can probably do a little better than this. In the first place Irving Langmuir's criteria (p. 79) have a quite remarkable generality, and the improbable nature of the claims made for the existence of N-rays, for instance, or of polywater, was undoubtedly apparent to the majority of informed observers at the time. Some hesitated to make the inescapable inference, because of the reputation of the protagonists: could someone with the record of professional competence and even high achievement of a Blondlot or a Derjaguin fall prey to such feckless credulity? The accusation would have required powerful confidence in one's judgement and perhaps a willingness to risk losing friends.

In general the adherents of the aberrant theory accept the inevitable when they realise that time and effort have not been rewarded by more convincing evidence; the dearth of proof becomes ever more embarrassing, and soon it seems that what reliable data there are actually contradict the theory. As the end approaches the enterprise comes to resemble a fatally holed battleship, its guns, manned by the few diehards who have not abandoned ship, still firing vainly on the encircling enemies, as it sinks from sight.

It has been known for the odd renegade in a hopeless cause to resort to outright fraud (not, so far as is known, a component in any of the stories related here) when casuistry fails. If this happens, it is not merely a matter of 'improving' or massaging data—a familiar enough transgression when a research worker thinks he has the answer; to take one example, the physicist Robert A. Millikan, founder President of the California Institute of Technology, who won the Nobel Prize for measuring the charge of the electron, appears to have been guilty of such a lapse. His notebooks, examined long after his death, revealed that he had thought it fitting to jettison deviant data points (trimming, in Charles Babbage's litany of vices); he had however good reasons for confidence in his conclusions and his published results have stood the test of time. Such proceedings of course are reprehensible, but outright invention of data is something quite other. A case recently unearthed was that of John Heslop Harrison, a respected botanist and Fellow of the Royal Society. He sought proof for his conviction that plants survived in a kind of suspended animation under the ice-sheet through geological time; but, believing evidently that his theory was foolproof, he decided to dispense with the laborious process of gathering evidence, and instead simply planted alien species on a remote Scottish island and then brought them back in triumph.

The germ of a pathological episode is usually an innocent mistake or an experimental mirage; the perpetrator is persuaded that he has made a great discovery, which will bring him fame and advancement in his profession. Once committed it is difficult to go back and to allow the principles of caution and scepticism that training and experience normally inculcate to overcome the excitement and euphoria of a brilliant success. Contrary evidence is brushed aside: *si incommodus est, non est.* The point of no return is soon reached: self-delusion, ambition, the damage to *amour propre* of a humiliating retraction, envy of colleagues or jealousy unbalance the intellect.

None of this explains how a mass-movement can begin: the urge to follow an admired or feared patron or torch-bearer may play its part. Awe of reputation or authority can of course paralyse the critical faculties, as can the prospect of a rapid rise in the scientific, even the social hierarchy. I am conscious that the histories I have related are of very diverse kinds, but they seem to me to be linked by a common thread. In all cases the aberration spreads like a contagion and is propagated by those who should have scotched it at the outset; and most often the participants include scientists of high repute, not noted for a frivolous acceptance of unproven, even anti-rational ideas; invariably perception is subjugated to desire.

One can, at the same time, discern differences in the underlying psychologies of these various episodes. The most common by far results from attempts to evaluate observations of a kind inherently difficult, or impossible to quantify by the techniques to hand. Does the clock-face on the other side of a darkened room become more distinct when one holds an iron bar above one's eyes? Does one midwife toad have more distinct swellings on its palms than another? An honest striving after objectivity is distorted by an inner conviction about the answer. Such was the genesis of N-rays, mitogenic radiation, Kammerer's Lamarckian experiments and many more. A second type of aberration arises from an almost wilful misinterpretation of data. A familiar parable goes like this: a zoologist sets out to find the relation between the number of an insect's legs and how far it can jump. He places it on the bench and claps his hands. The creature jumps, and he records the distance. He then cuts of one of its legs and repeats the experiment. The insect manages to jump until only one leg remains. The zoologist cuts off the last leg and claps his hands; the insect remains motionless. He claps again with the same result and then takes his notebook and writes: 'Removal of last leg destroys hearing'. The story of cold fusion has elements in which this kind of reasoning is evident. One can see how the theory about the insect's hearing could divide entomologists.

Next, is seeing only the wood, but not the trees: a seemingly remarkable new phenomenon is observed and the interesting interpretation takes precedence over the trivial. The possibility of epiphenomena, of unperceived experimental imperfections, is simply not entertained. In this way the validity of the supposed phenomenon is taken on trust. Thus the principle (akin to Hume's dictum on miracles (p. 97)) that the more probable explanation should be thoroughly explored before the improbable is contemplated, is recklessly violated. Inadequate experimental procedures, especially those that omit essential controls, form another category: dropped and floating organs are an example, as are the episodes described with such relish by Irving Langmuir on electron capture and the Faraday effect. Many of the studies on supposed paranormal phenomena suffered from the same deficiencies of experimental design, but one could regard credulity in simply accepting the word of hoaxers and charlatans as forming a class all to itself. And finally there is the insidious influence of political ideology, which pervaded eugenics and the sinister pseudo-science of Lysenko and his followers and of the leaders of the Aryan biology, physics and mathematics movements in the Third Reich. There were undoubtedly many who subscribed to these schools for purely cynical reasons of personal advancement, but of far greater interest are those (often able scientists) who evidently perceived an intellectual imperative in such bizarre and pernicious doctrines. In the West, even such patrician scholars as J.D. Bernal and J.B.S. Haldane were slow to condemn the destructive folly of what was happening in the Soviet Union during Lysenko's reign.

So we return to the conclusion—and who would question it?—that scientists, for all their vaunted training in observation and scepticism, are as much a prey to human frailty as anyone else, and their capacity for unbending objectivity is circumscribed. Even scepticism has its dark side. Schopenhauer wrote that all truth passes through three stages: first it is ridiculed, then violently opposed and finally accepted as self-evident. Charles Kettering, the mogul of General Motors, enlarged on this enduring article of human nature: 'First they tell you you're wrong and they can prove it; then they tell you you're right but it isn't important; then they tell you it's important but they knew it all along'.

Further Reading

Suggested books or articles which may be of interest to the general reader are indicated by asterisks; the remainder, some of them primary sources, are for the most part of more specialised interest.

Introduction

For accounts of fraud in science see **Betrayers of the Truth—Fraud and Deceit in the Halls of Science* by William J. Broad and Nicholas Wade (Simon and Schuster, New York, 1982); *False Prophets: Fraud and Error in Science and Medicine* by Alexander Kohn (Blackwell, Oxford, 1986); and A. Mazur, 'Allegations of dishonesty in research and their treatment in American universities', *Minerva*, **27**, 177 (1989).

For fraud in medical research there is *Fraud and Misconduct in Medical Research* by S.J. Lock and F.O. Wells (BMJ Publishing Group, London, 1996).

As a survey of the lunatic fringe, Martin Gardner's book **In the Name of Science*, expanded in a second edition under the title **Fads and Fallacies in the Name of Science* (Dover, New York, 1957), remains a classic. Also indispensable is **An Encyclopedia of Claims, Frauds and Hoaxes of the Occult and Supernatural* by James Randi (St. Martin's Press, New York, 1995).

Chapter 1: Blondlot and the N-rays

Entertaining accounts of the N-ray affair can be found in Irving M. Klotz's book, **Diamond Dealers and Feather Merchants* (Birkhäuser, Boston, Basel and Stuttgart, 1985) and in his article in the **Scientific American*, **243**, 122 (May 1980). A selection of Blondlot's papers on aspects of N-rays from the *Comptes rendus de l'Académie des Sciences* appeared in an English translation by J. Garcin as *'N' Rays—A Collection of Papers Communicated to the Academy of Sciences by*

R. Blondlot (Longmans Green and Co., London, 1905), with a brief commentary by the translator. Note that the publication date is the year following R.W. Wood's devastating demolition of the N-rays in *Nature*, **70**, 530 (1904).

For a scholarly study of N-rays and the historical context see: Mary-Jo Nye, *Historical Studies in the Physical Sciences*, **11**, 125 (1981), as well as her book, *Science in the Provinces—Scientific Communities and Provincial Leadership in France, 1860–1930* (University of California Press, Berkeley and London, 1986) for the general social background.

Chapter 2: Paradigms Enow: Some Mirages of Biology

I have been able to find no history of mitogenic radiation. The book by Anna Gurvich (Gurwitsch) on the subject is in German (*Die Mitogenetische Strahlung—Probleme der Zellteilung*, Springer, Berlin, 1932); it gives a stubbornly dogmatic, very detailed account of the results, all of them positive, when the existence of the phenomenon was already being widely questioned. The following is a selection from the many papers on mitogenic radiation phenomena in biological and chemical systems. I have chosen only from those which appeared in the more highly ranked journals: A. Potozky and A.E. Braunstein, *Biochemische Zeitschrift*, **249**, 270, 282 (1932); J. Klenitzky, *Biochemische Zeitschrift*, **252**, 126 (1932); M. Heinemann, *Nature*, **134**, 701 (1934); J.B. Bateman, *Biological Reviews*, **10**, 42 (1935); W.W. Siebert and H. Seffert (who led in finding startling differences in radiations from subjects afflicted with anything from lassitude to cancer), *Biochemische Zeitschrift*, **287**, 104, 109 (1936) and **301**, 301 (1939). Perhaps the most incisive studies of the phenomenon, which should have persuaded the scientific world that there were no mitogenic rays, were those by E. Lorenz, *Physical Reviews*, **44**, 329 (1933) and by J. Gray and C. Ouellet, *Proceedings of the Royal Society B*, **114**, 1 (1933). In a late attempt to find a theoretical justification for his radiation, Gurvich enlisted the help of a distinguished physicist: Ya.I. Frenkel and A.G. Gurvich, *Transactions of the Faraday Society*, **39**, 201 (1943). Another believer was Otto Rahn, who published his book *Invisible Radiation of Organisms* (Borntraeger, Berlin) in 1936. For an account of Gurvich's life and some of his work by his grandson and disciple, see L.V. Beloussov, *Journal of Developmental Biology*, **41**, 771 (1997); see also the late review by Gurvich's daughter: A.A. Gurwitsch, *Experientia*, **44**, 545 (1992). An attempt to rehabilitate mitogenic radiation is by M. Wainwright, *Perspectives in Biology and Medicine*, **41**, 565 (1998).

The story of Abderhalden and his elusive enzymes is related in an absorbing article by Ute Deichmann and Benno Müller-Hill: *'The fraud of Abderhalden's enzymes', *Nature*, **393**, 109 (1998). For anyone wanting to enter into greater detail or to catch the tone of the debate between Abderhalden and his critics, I give a few of the more prominent papers in what are (or were at the time) well-regarded journals with a wide circulation: L. Michaelis and L. von Langemark, *Deutsche medizinische Wochenschrift*, **40**, 317 (1914), followed by Abderhalden's attempted rebuttal and a rejoinder by Michaelis; D.D. van Slyke and M. Vinograd, *Proceedings of the Society of Experimental Biology and Medicine*, **11**, 317 (1914); H. Singer, *Münchener medizinische Wochenschrift*, **61**, 350 (1914); W.N. Boldyrev, *Comptes rendus de l'Académie des Sciences*, **80**, 882 (1917).

Arthur Koestler's book, **The Case of the Midwife Toad* (Random House, New York, 1972) is a characteristically readable and wrongheaded version of Paul Kammerer's rise and fall.

For Ungar's memory transfer experiments, see the paper 'Isolation, identification and synthesis of specific-behaviour-inducing brain peptide' by G. Ungar, D.M. Desiderio and W. Parr, *Nature*, **238**, 198 (1972), and the critical analysis of the data by the reviewer, W.W. Stewart, which follows it (p. 202). Among the weightiest publications from other laboratories, purportedly confirming the existance of memory transfer are: F. Rosenblatt, J.T. Farrow and W.F. Herblin, *Nature*, **209**, 46 (1966); F.R. Babich *et al.*, *Science*, **149**, 652 (1965); A.L. Jacobson *et al.*, *Science*, **150**, 636 (1965). For reports of learning-related changes in brain RNA see H. Hydén and E. Egyhazi, *Proceedings of the National Academy of Sciences of the U.S.A.*, **49**, 618

(1963); **52**, 1030 (1964). Papers reporting absence of any memory transfer effects are by C.G. Gross and F.M. Carey, *Science*, **150**, 1749 (1965) and M. Lutthers *et al.*, *Science*, **151**, 835 (1966).

Chapter 3: Aberrations of Physics: Irving Langmuir Investigates

The interested reader can do no better than go to Irving Langmuir's highly personal account, entitled 'Pathological Science'. The transcript of this lecture, given in 1953, is reprinted in **Physics Today*, **42**, 36 (1989).

The primary original papers on the electron capture experiments were: B. Davis and A. Barnes, *Physical Reviews*, **34**, 152 (1929) and A. Barnes, *Physical Reviews*, **35**, 217 (1930). The first report of failure to reproduce the effect was by H.C. Webster, *Nature*, **126**, 352 (1930).

The literature on the Allison effect is extensive. The original description of the imaginary phenomenon is given by F. Allison and E.S. Murphy, *Journal of the American Chemical Society*, **52**, 3796 (1930). For the supposed discovery of numerous isotopes see, for example, W.M. Latimer and H.A. Young, *Physical Reviews*, **44**, 690 (1933) and L.B. Snoddy, *Physical Reviews*, **44**, 691 (1933); resounding successes in identifying unknown samples are reported by T.R. Ball, *Physical Reviews*, **47**, 548 (1935). A highly sanguine survey of the Allison effect, running to three hefty articles and admitting of no possible doubts about its validity, is by S.S. Cooper and T.R. Ball, *Journal of Chemical Education*, **13**, 210, 278 and 326 (1936). Among those who obtained only negative results were H.G. MacPherson, *Physical Reviews*, **47**, 310 (1935); G. Hughes, *Journal of the American Chemical Society*, **58**, 1924 (1936) and G.C. Comstock, *Physical Reviews*, **51**, 776 (1937); but the same year D.C. Bond reported in the *Journal of the American Chemical Society*, **59**, 439 (1937) that the Allison effect was 'definitely real'.

Chapter 4: Nor Any Drop to Drink: the Tale of Polywater

The most comprehensive and authoritative account of the polywater episode is contained in the book **Polywater* by Felix Franks (MIT Press, Cambridge, Massachusetts, 1981). Some of the key articles in the exposure of the artefact are: W.D. Bascom, E.J. Brooks and B.N. Worthington, *Nature*, **228**, 1290 (1970); D.L. Rousseau and S.P.S. Porto, *Science*, **167**, 1715 (1970); S.L. Kurtin *et al.*, *Science*, **167**, 1720 (1970); D.H. Everett, J.M. Haynes and P.J. McElroy, *Nature*, **226**, 1033 (1970); W.M. Madigosky, *Science*, **172**, 259 (1971).

Chapter 5: The Wilder Shores of Credulity

For the story of Andrew Crosse's career and its Victorian setting, see the chapter by J.A. Secord, 'Extraordinary experiment: electricity and the creation of life in Victorian England' in *The Uses of Experiment—Studies in the Natural Sciences* (D. Gooding, T. Pinch and S. Schaffer, eds, Cambridge University Press, 1989), p. 337.

For the history of parapsychology and related matters up to about 1955 one can still do no better than to consult Martin Gardner's sprightly classic, **Fads and Fallacies in the Name of Science* (Dover, New York, 1957). This is a revised edition of the earlier book *In the Name of Science.* Later escapades are anatomised in some of Martin Gardner's articles, collected in **Science Good, Bad and Bogus* (Oxford University Press, Oxford, 1981). James Randi's study, **The Magic of Uri Geller* (Ballantine Books, New York, 1975), should also be consulted. The extraordinary spirit of credulity with which a group of academics in Britain approached the investigation of spoon-bending and related pseudo-phenomena is alarmingly encapsulated in John G. Taylor's *Superminds* (Macmillan, London, 1975). A good survey of these odd events is set out in an editorial in **Nature*, **254**, 470 (1975); a description of the antics of the children when left alone with the cutlery is to be found in an article by B.R. Pamplin and H. Collins, *Nature*, **257**, 8 (1975), and on the same page there is an acerbic comment by James Randi, with the title: 'Presto! Would you believe?' Taylor progressively distanced himself from his earlier assertions in two articles: E. Balanovsky and J.G. Taylor, *Nature*, **276**, 64 (1978) and J.G. Taylor and E. Balanovsky, *Nature*, **279**, 631 (1979). The second bears the title: 'Is there any scientific explanation of the paranormal?', to which the answer was 'no'.

Chapter 6: Energy Unlimited

There are three first-rate books to choose from on cold fusion. The most compelling, to my mind, is Gary Taubes's **Bad Science—The Short Life and Weird Times of Cold Fusion* (Random House, New York, 1993). The others are **Too Hot to Handle: the Race for Cold Fusion* by Frank Close (W.H. Allen, London, 1990) and **Cold Fusion. The Scientific Fiasco of the Century* (Oxford University Press, Oxford, 1993) by John R. Huizenga, who was one of the experts called in to investigate the claims.

Chapter 7: What the Doctor Ordered

An absorbing history of unnecessary surgery is to be found in **Fantasy Surgery 1880–1930* by Ann Dally (Rodopi, Amsterdam and Atlanta, Georgia, 1996). For a starry-eyed biography of Sir William Arbuthnot Lane see T.B. Layton: *W.A. Lane* (Livingstone, Edinburgh and London, 1956).

The story of Voronoff and of testis implants and related treatments is told in **The Monkey Gland Affair* by David Hamilton (Chatto and Windus, London, 1986), an account that can scarcely be bettered. Élie Metchnikoff's somewhat deranged book, *The Nature of Man: Studies in Optimistic Philosophy*, translated by P. Chalmers Mitchell (W. Heinemann, London, 1905), stands as an odd period-piece.

For a grisly case of misuse of radioactivity see Roger M. Macklis, *Scientific American*, **269**, 94 (August 1993).

Chapter 8: Science, Chauvinism and Bigotry

Pierre Duhem's lectures, translated under the title **German Science* (Open Court, La Salle, Illinois, 1991), convey the flavour of academic discourse in a Europe riven by war. On the attitudes of French scientists to their German brethren in the years up to the end of the Great War, see also the short monograph by Harry W. Paul, *The Sorcerer's Apprentice—The French Scientist's Image of German Science 1840–1919* (University of Florida Press, Gainesville, Florida, 1972). For the chauvinistic excesses of German chemists, particularly Kolbe, see A.J. Rocke, *Ambix*, **34**, 156 (1987), while P. Forman, *Isis*, **64**, 151 (1971) expatiates on the attitudes of physicists. An authoritative account of the nationalistic animosities that divided the French and German scientific communities is 'Von der Science Allemande zur Deutschen Physik' by A. Kleinert, *Francia*, **6**, 509 (1978).

Chapter 9: A Climate of Fear

The literature on science in the Soviet Union is vast. On Lysenko and his influence there are three books, all well worth reading. **The Rise and Fall of T.D. Lysenko* (Columbia University Press, New York, 1969) is by Zhores A. Medvedev, who suffered through the period; David Joravsky's **The Lysenko Affair* (Harvard University Press, 1970) is the most scholarly study, and Valery N. Soyfer's **Lysenko and the Tragedy of Soviet Science* (Rutgers University Press, 1994) is also interesting, as is a brief account by the same author in *Nature*, **339**, 415 (1989). For a survey of human genetics in the USSR after the Revolution and the events leading up to Lysenko's ascendency, see also M.B. Adams, *Genome*, **31**, 879 (1989). The impact of Lysenkoism in Poland is recounted in an interesting article by W. Gajewski, under the title, 'The grim heritage of Lysenkoism. II. Lysenkoism in Poland' in **Quarterly Reviews of Biology*, **65**, 423 (1990).

More general histories of Soviet science and Soviet thought are by the foremost Western scholar of the period, Loren R. Graham, and include **Science and Philosophy in the Soviet Union* (A.A. Knopf, New York, 1971) and *Science in Russia and the Soviet Union: A Short History* (Columbia University Press, New York, 1993). Another authoritative work is again by David Joravsky: **Soviet Marxism and Natural Science 1917–1932* (Columbia University Press, New York, 1961). Also recommended are **Manipulated Science—The Crisis of Science and Scientists in the Soviet Union Today*, by a Russian journalist, Mark Popovsky, translated by Paul S. Falla (Doubleday, New York, 1979); **Soviet Science* by Z.A. Medvedev (W.W. Norton, New York, 1978); **Stalinist Science* by Nikolai Krementsov (Princeton University Press, Princeton, 1997), whose article on the 'K-R affair', *Historical*

Studies in the Life Sciences, **17**, 419 (1995) is also of interest; *Soviet Science under Control—The Struggle for Influence*, by J.L. Roberg (Macmillan Press, London and St. Martin's Press, New York, 1998).

An excellent account of what the physicists had to endure in creating the Soviet atomic bomb is David Holloway's **Stalin and the Bomb—The Soviet Union and Atomic Energy* (Yale University Press, New Haven, 1994); and see also Richard Rhodes: **Dark Sun: the Making of the Hydrogen Bomb* (Scribners, New York, 1995).

For the débâcle of resonance theory in the Soviet Union see the interesting article by István Hargittai, **Chemical Intelligencer*, **1**, 34 (1995).

Chapter 10: Science in the Third Reich: Bigotry, Racism and Extinction

A large proportion of the abundant literature on all aspects of science in the Third Reich is inevitably in German. The best general account in English is A.D. Beyerchen's **Scientists under Hitler* (Yale University Press, New Haven, 1977) (and see also Beyerchen's article in *Social Research*, **59**, 615 (1992)). There is wide coverage in a multi-author volume, edited by M. Renneberg and M. Walker, **Science, Technology and National Socialism* (Cambridge University Press, Cambridge, 1994). This contains articles by historians on most aspects of science during the period. Also recommended is Ute Deichmann's **Biologists under Hitler* (Harvard University Press, Cambridge, Massachusetts, 1996). Among articles focusing on Nazi party science policy and the reactions to it of the professional bodies are W. Schlicker and J. Glaser, *Zeitschrift für Geschichte der Wissenschaft*, **31**, 881 (1983) and F. Graf-Schulhofer, *Zeitschrift für Geschichte der Wissenschaft*, **44**, 143 (1996) (both in German); see also *Universität unterm Hakenkreuz* by Helmut Heiber (K.E. Saur, Munich, 1991); Part 3 (1933–1945) of *Die Berliner Akademie der Wissenschaften in der Zeit des Imperialismus* (C. Grau, W.Schlicker and L. Zeil, ed., Akademie-Verlag, Berlin, 1979) and articles in *Hochschule und Nazionalsozialismus* (L. Siegele-Wenschelkreutz and G. Stuchlik, eds, Haag und Herchen Verlag,

Frankfurt-am-Main, 1990). *Surviving the Swastika* by Kristie Macrakis (Oxford University Press, Oxford and New York, 1993) details the history of the Kaiser-Wilhelm-Gesellschaft during the Third Reich.

I have refrained, so far as possible, from giving German titles as the main sources, but the only study of the Ahnenerbe, and one that makes engrossing reading, remains unaccountably untranslated: this is **Das 'Ahnenerbe' der SS 1935–1945* by M.H. Kater (Deutsche Verlagsanstalt, Stuttgart, 1974). A rather discursive account of the bizarre *Welteislehre* is to be found in *Universal Ice—Science and Ideology in the Nazi State*, by Robert Bowen (Belhaven Press, London, 1993). More scholarly is the book (in German) by Brigitte Nagel: **Die Welteislehre. Ihre Geschichte und ihre Rolle im 'Dritten Reich'* (Verlag für Geschichte der Naturwissenschaften und Technik, Stuttgart, 1991). The leader of the Alsos mission, Sam Goudsmit, wrote an entertaining chronicle of his discoveries in Germany in 1945. His book is called **Alsos* (Sigma Books, London, and H. Schuman, New York, 1947). For the background of the struggle of the physics establishment against the *Deutsche Physik* see J.L. Heilbron's biography of Max Planck, **The Dilemmas of an Upright Man* (University of California Press, Berkeley, Los Angeles, London, 1986); the mental turmoils suffered by many physicists when their subject underwent the upheavals of the new physics during the first decade of the twentieth century are movingly described in fictionalised form by Russell McCormmach in **Night Thoughts of a Classical Physicist* (Harvard University Press, Cambridge, Massachusetts, 1982).

For more specialised studies on aspects of physics in the Third Reich, see S. Richter, *Sudhoffs Archiv*, **57**, 195 (1973), A. Kleinert on the exchange of letters between Lenard and Stark, especially in regard to control of the Kaiser-Wilhelm-Gesellschaft in *Physikalische Blätter*, **36**, 35 (1980), and Joachim Stark's unrepentant and distorted, though curiously mesmeric memoirs, *Erinnerungen eines deutschen Naturforschers* (A. Kleinert, ed., Bionamica Verlag, Mannheim, 1987) and his angry response to criticisms in *Physikalische Blätter*, **3**, 271

(1947), followed by a dignified rejoinder by Max von Laue (all in German); the story of the tribunal proceedings against Stark after the end of the war is told by A. Kleinert, *Sudhoffs Archiv*, **67**, 13 (1983).

On *Deutsche Mathematik* and its origins the articles by D.E. Rowe, **Isis*, **77**, 429 (1986) and A. Shields, **Mathematical Intelligencer*, **3**, 7 (1988) contain much of interest. There are also relevant chapters in a volume in French, entitled *La Science dans le Troisième Reich* (Josiane Olff-Nathan, ed., Éditions du Seuil, Paris, 1993). G.H. Hardy's observations on Bieberbach's racial opinions are in *Nature*, **134**, 250 (1934).

An excellent treatise on biology under the Nazi regime is Änne Bäumer's **NS-Biologie* (S. Hirzel—Wissenschaftliche Verlagsgesellschaft, Stuttgart, 1990), and the same author (Änne Bäumer-Schleinkofer) has written on biological education in state schools during this period: *NS-Biologie und Schule* (Peter Lang, Frankfurt-am-Main, 1992). On Nazi medicine there is a wide choice of reading. A good start would be **Murderous Science—Elimination by Selection of Jews, Gypsies and Others, Germany 1933–1945* (Oxford University Press, Oxford, 1988) by Benno Müller-Hill; and Robert N. Proctor's **Racial Hygiene—Medicine under the Nazis* (Harvard University Press, Cambridge, Massachusetts, 1988). Several informative articles on medicine in the Third Reich can be found in *Medizin, Naturwissenschaft, Technik und Nazionalsozialismus—Kontinuitäten und Diskontinuitäten* (C. Meinel and P. Voswinckel, eds, Verlag für Geschichte der Naturwissenschaften und Technik, Stuttgart, 1994); this volume also contains an article on Conrad Weygand and the *Deutsche Chemie*.

For the unexpectedly progressive side of social medicine in Nazi Germany see Robert N. Proctor's **The Nazi War on Cancer* (Princeton University Press, 1999).

Chapter 11: Nature Nutured: the Rise and Fall of Eugenics

As a comprehensive, authoritative and readable history of the subject of eugenics, **In the Name of Eugenics: Genetics and the Uses of Human*

Heredity, 2nd edition, by Daniel J. Kevles (Harvard University Press, Cambridge, Massachusetts, 1995, and Penguin Books, Harmondsworth), stands alone. Also recommended is **The Nazi Connection—Eugenics, American Racism and German National Socialism*, by Stefan Kühl (Oxford University Press, Oxford, 1994), and there is interesting material in Paul R. Josephson's book, *Totalitarian Science and Technology* (Humanistic Press, New Jersey, 1996). A good article on the British scene at the beginning of the twentieth century is by E.J. Larson, *British Journal for the History of Science*, **24**, 45 (1991). On the debate about the genetics of intelligence, Stephen J. Gould's **The Mismeasure of Man* (W.W. Norton, New York, 1981) is also relevant.

Index